Association for Women in Mathematics Series

Volume 22

Series Editor
Kristin Lauter
Microsoft Research
Redmond, WA, USA

Association for Women in Mathematics Series

Focusing on the groundbreaking work of women in mathematics past, present, and future, Springer's Association for Women in Mathematics Series presents the latest research and proceedings of conferences worldwide organized by the Association for Women in Mathematics (AWM). All works are peer-reviewed to meet the highest standards of scientific literature, while presenting topics at the cutting edge of pure and applied mathematics, as well as in the areas of mathematical education and history. Since its inception in 1971, The Association for Women in Mathematics has been a non-profit organization designed to help encourage women and girls to study and pursue active careers in mathematics and the mathematical sciences and to promote equal opportunity and equal treatment of women and girls in the mathematical sciences. Currently, the organization represents more than 3000 members and 200 institutions constituting a broad spectrum of the mathematical community in the United States and around the world.

Titles from this series are indexed by Scopus.

More information about this series at http://www.springer.com/series/13764

Rebecca Segal • Blerta Shtylla • Suzanne Sindi
Editors

Using Mathematics to Understand Biological Complexity

From Cells to Populations

 Springer

Editors
Rebecca Segal
Department of Mathematics
and Applied Mathematics
Virginia Commonwealth University
Richmond, VA, USA

Blerta Shtylla
Department of Mathematics
Pomona College
Claremont, CA, USA

Suzanne Sindi
Department of Applied Mathematics
University of California
Merced, CA, USA

ISSN 2364-5733   ISSN 2364-5741 (electronic)
Association for Women in Mathematics Series
ISBN 978-3-030-57131-3   ISBN 978-3-030-57129-0 (eBook)
https://doi.org/10.1007/978-3-030-57129-0

Mathematics Subject Classification (2020): 92-06, 92-B05

This Springer imprint is published by the registered company Springer Nature Switzerland AG
The registered company address is: Gewerbestrasse 11, 6330 Cham, Switzerland

Contents

Collaborative Workshop for Women in Mathematical Biology

Rebecca Segal, Blerta Shtylla, and Suzanne Sindi

1 Aim and Scope

Biological systems are complex and highly interconnected. Despite increasing amounts of information collected, it is not always clear how to use these data to make conclusions and predictions. Mathematical models are powerful tools in biology because they allow us to abstract the biological system in order to frame questions, explore patterns and synthesize information. Indeed, we are writing these remarks during the COVID-19 Pandemic which has illustrated in a staggering way the importance of quantitative modeling in aiding our understanding of complex biological processes. This volume contains the scientific and collaborative work from the *Collaborative Workshop for Women in Mathematical Biology*. The workshop brought together forty-five researchers to collaborate on seven problems each of which used mathematics to understand complex biological systems. The workshop was held at the Institute of Pure and Applied Mathematics on the campus of University of California, Los Angeles from June 17-21, 2019 in Los Angeles, CA and was organized by Rebecca Segal, Blerta Shtylla, and Suzanne Sindi. The articles

R. Segal (✉)
Department of Mathematics and Applied Mathematics, Virginia Commonwealth University, Richmond, VA, USA
e-mail: rasegal@vcu.edu

B. Shtylla
Department of Mathematics, Pomona College, Claremont, CA, USA
e-mail: shtyllab@pomona.edu

S. Sindi
Department of Applied Mathematics, University of California, Merced, CA, USA
e-mail: ssindi@ucmerced.edu

© The Association for Women in Mathematics and the Author(s) 2021
R. Segal et al. (eds.), *Using Mathematics to Understand Biological Complexity*,
Association for Women in Mathematics Series 22,
https://doi.org/10.1007/978-3-030-57129-0_1

1

contained in this volume were initiated during the intensive one-week workshop and continued through follow-up collaborations afterwards.

2 History and Context

Historically, women have been underrepresented in the mathematical sciences. Although progress has been made, the numbers remain unbalanced. In the most recent American Mathematical Society Survey from 2017, only 17% of tenure-track mathematics faculty in doctoral departments are female (http://www.ams.org/profession/data/annual-survey/demographics). A specific breakdown of distribution within different types of institutions (https://www.womendomath.org/research/) gives an even more compelling picture of why research workshops such as this one can be so valuable for the mathematics community. Research mentoring and support from senior mentors is one key to success and a workshop environment provides a significant amount of interaction in a concentrated amount of time.

The primary aim of the Women in Mathematical Biology (WIMB) workshops is to foster research collaboration among women in mathematical biology. Participants spend a week making progress on a research project and encouraging innovation in the application of mathematical, statistical, and computational methods in the resolution of significant problems in the biosciences. The workshops have a special format designed to maximize the opportunities to collaborate. The groups are structured to facilitate tiered mentoring. Each group has a senior researcher who presents a problem. This person is matched with a co-leader, typically a researcher in their field but with whom they have not previously collaborated. The groups are rounded out with researchers at various career stages. By matching senior research mentors with junior mathematicians, we expand and support the community of scholars in the mathematical biosciences. To date, WIMB workshops have occurred at the Institute for Mathematics and its Applications (IMA, https://www.ima.umn.edu/), the National Institute for Mathematical and Biological Synthesis (NIMBioS, http://www.nimbios.org/), the Mathematical Biosciences Institute (MBI,https://mbi.osu.edu/), and most recently at the Institute of Pure and Applied Mathematics (IPAM, https://www.ipam.ucla.edu/). These workshops have been sponsored by an ADVANCE grant from the National Science Foundation to the Association for Women in Mathematics. This award has helped establish research networks in 21 different areas of mathematics research including Control, Commutative Algebra, Geometry, Data Science, Materials, Operator Algebras, Analysis, Number Theory, Shape, Topology, Numerical Analysis, and Representation Theory.

For the Mathematical Biology workshops, each group continues their project together to obtain results that are submitted to the peer-reviewed AWM Proceedings volume for the workshop. The benefit of such a structured program with leaders, projects and working groups planned in advance is based on the successful Women In Numbers (WIN) conferences and works in both directions: senior women will meet, mentor, and collaborate with the brightest young women in their field on a part of their research agenda of their choosing, and junior women and students will develop their network of colleagues and supporters and encounter important new research areas to work in, thereby improving their chances for successful research careers.

One of the most critical workshop goals is help establish supportive and productive research groups that are sustained well beyond the workshop. Below we include some representative statements from participants shared with us when we surveyed them at the end of the workshop to assess their opinion of the workshop structure and the impact of the workshop on their careers. The group dynamics were overwhelmingly listed as a positive experience: *"The opportunity to work with, share ideas, and learn from a group made up entirely of female mathematicians."* Some participants appreciated the format of the workshop for allowing *"Exclusive time spent working with talented people on a new project."* Participants left the workshop satisfied with their experience: *"Establishing a new group of collaborators. I've honestly never developed this skill and I'm glad to have had this opportunity."* The workshop sometimes stretched participants out of their comfort zone while still providing a positive experience: *"Watching in awe as phenomenal women worked on math and bio. I tried my best to contribute, and although I felt like I still lacked a lot of background to really make a real impact, it was really inspirational to learn from women established in their careers. I definitely have a lot more role models at the end of this trip! The industry panel was helpful in showing me more career opportunities for a mathematical biologist."* The group leaders were pleased with the work produced by their teams and all of the participants learned new mathematics, new biology, or new computational tools to move the research project in a productive direction. Finally, as organizers we have been delighted to see the teams initiated at these workshops produce new research projects, papers, proposals and other scholarly products far beyond the scope of the original team lead project.

3 Research

Within this volume are mathematical research papers covering a wide range of application areas. The work can be loosely grouped into a few general application areas: structural organization of biological material, infection modeling, and disease treatment. Throughout this research are discussions of how to create accurate models with limited data, how to work across biological scales, and how to best describe complex structures in a useful manner.

Several teams had research related to the structural organization of organisms. One project focused on how the protein actin helps form larger structures within a cell. Other projects studied DNA topology and DNA secondary structure to understand the design and replication mechanisms in organisms. Actin assembles into semi-flexible filaments that cross-link to form higher order structures within the cytoskeleton. This study focused on the dynamics of the formation of a branched actin structure as observed at the leading edge of motile eukaryotic cells. They constructed a minimal agent-based model for the microscale branching actin dynamics, and a deterministic partial differential equation model for the macroscopic network growth and bulk diffusion. Their results suggest that perturbations to microscale rates can have significant consequences at the macroscopic level, and these should be taken into account when proposing continuum models of actin network dynamics.

DNA topology, formal grammar R-loops, are three-stranded nucleic acid structures consisting of two DNA strands and one RNA strand. They form naturally during transcription when the nascent RNA hybridizes to the template DNA strand, forcing the coding DNA strand to wrap around the RNA:DNA duplex. In their study, this team used words generated by the grammars to represent topological segments of the DNA:DNA and RNA:DNA interactions. They extended this model to include properties of the DNA nucleotide sequence.

A third group explored the extent to which graph algorithms for community detection can improve the mining of structural information from the predicted Boltzmann/Gibbs ensemble for the biological objects known as RNA secondary structures. Since more structural information is obtained in 50% of the test cases, this proof-of-principle analysis supports efforts to improve the data mining of RNA secondary structure ensembles.

Two groups worked broadly in the area of infection: one group examined disease spread across geographic regions while another group explored in host resolution of infection. How do interventions impact malaria dynamics between neighboring countries? Although many countries world wide have taken measures to decrease the incidence of malaria many regions remain endemic, and in some parts of the world malaria incidence is increasing. This team considered the case of neighboring countries Botswana and Zimbabwe, connected by human mobility. They used a two-patch Ross-MacDonald Model with Lagrangian human mobility to examine the coupled disease dynamics between these two countries.

Antimicrobial resistance (AMR) is a serious threat to global health today. The spread of AMR, along with the lack of new drug classes in the antibiotic pipeline, has resulted in a renewed interest in phage therapy, which is the use of bacteriophages to treat pathogenic bacterial infections. These researchers utilized a mathematical model to examine the role of the immune response in concert with phage-antibiotic combination therapy compounded with the effects of the immune system on the phages being used for treatment. They explored the effect of phage-antibiotic combination therapy by adjusting the phage and antibiotics dose or altering the timing. Their results show that it is important to consider the host immune system in mathematical models and that frequency and dose of treatment are important considerations for the effectiveness of treatment.

Finally, two groups worked broadly in the area of disease progression and treatment. One group developed a model for retinal degeneration while the other focused on radiation therapy for cancerous tumors. In the retina, photoreceptor degeneration can result from imbalance in lactate production and consumption as well as disturbances to pyruvate and glucose levels. To identify the key mechanisms in metabolism that may be culprits of this degeneration, they used a nonlinear system of differential equations to mathematically model the metabolic pathway of aerobic glycolysis in a healthy cone photoreceptor. Their model allowed them to analyze the levels of lactate, glucose, and pyruvate within a single cone cell. They performed numerical simulations, used available metabolic data to estimate parameters and fit the model to this data, and conducted a sensitivity analysis using two different methods (LHS/PRCC and eFAST) to identify pathways that have the largest impact on the system.

Recent technological advances make it possible to collect detailed information about tumors, and yet clinical assessments about treatment responses are typically based on sparse datasets. In this work, one team proposed a workflow for choosing an appropriate model, verifying parameter identifiability, and assessing the amount of data necessary to accurately calibrate model parameters. They considered a simple, one-compartment ordinary differential equation model which tracks tumor volume and a two-compartment model that accounts for tumor volume and the fraction of necrotic cells contained within the tumor.

4 Concluding Remarks

It merits note that the majority of revisions for this volume were accomplished during the COVID-19 pandemic; we are both grateful for and proud of the hard work

Fig. 1 Group photograph of the workshop participants

Connecting Actin Polymer Dynamics Across Multiple Scales

Calina Copos, Brittany Bannish, Kelsey Gasior, Rebecca L. Pinals, Minghao W. Rostami, and Adriana T. Dawes

Abstract Actin is an intracellular protein that constitutes a primary component of the cellular cytoskeleton and is accordingly crucial for various cell functions. Actin assembles into semi-flexible filaments that cross-link to form higher order structures within the cytoskeleton. In turn, the actin cytoskeleton regulates cell shape, and participates in cell migration and division. A variety of theoretical models have been proposed to investigate actin dynamics across distinct scales, from the stochastic nature of protein and molecular motor dynamics to the deterministic macroscopic behavior of the cytoskeleton. Yet, the relationship between molecular-level actin processes and cellular-level actin network behavior remains understudied, where prior models do not holistically bridge the two scales together.

In this work, we focus on the dynamics of the formation of a branched actin structure as observed at the leading edge of motile eukaryotic cells. We construct

The authors "Brittany Bannish, Kelsey Gasior, Rebecca L. Pinals, and Minghao W. Rostami" contributed equally.

C. Copos
Department of Mathematics and Computational Medicine Program, University of North Carolina Chapel Hill, Chapel Hill, NC, USA

B. Bannish
Department of Mathematics & Statistics, University of Central Oklahoma, Edmond, OK, USA

K. Gasior
Department of Mathematics, Florida State University, Tallahassee, FL, USA

R. L. Pinals
Department of Chemical and Biomolecular Engineering, University of California, Berkeley, CA, USA

M. W. Rostami
Department of Mathematics and BioInspired Syracuse, Syracuse University, Syracuse, NY, USA

A. T. Dawes (✉)
The Ohio State University, Columbus, OH, USA
e-mail: dawes.33@osu.edu

© The Association for Women in Mathematics and the Author(s) 2021
R. Segal et al. (eds.), *Using Mathematics to Understand Biological Complexity*, Association for Women in Mathematics Series 22, https://doi.org/10.1007/978-3-030-57129-0_2

a minimal agent-based model for the microscale branching actin dynamics, and a deterministic partial differential equation (PDE) model for the macroscopic network growth and bulk diffusion. The microscale model is stochastic, as its dynamics are based on molecular level effects. The effective diffusion constant and reaction rates of the deterministic model are calculated from averaged simulations of the microscale model, using the mean displacement of the network front and characteristics of the actin network density. With this method, we design concrete metrics that connect phenomenological parameters in the reaction-diffusion system to the biochemical molecular rates typically measured experimentally. A parameter sensitivity analysis in the stochastic agent-based model shows that the effective diffusion and growth constants vary with branching parameters in a complementary way to ensure that the outward speed of the network remains fixed. These results suggest that perturbations to microscale rates can have significant consequences at the macroscopic level, and these should be taken into account when proposing continuum models of actin network dynamics.

Keywords Actin · Differential equations · Stochastic model · Sensitivity analysis · Cytoskeleton

1 Introduction

A cell's mechanical properties are determined by the cytoskeleton whose primary components are actin filaments (F-actin) [1–4]. Actin filaments are linear polymers of the abundant intracellular protein actin [5–7], referred to as G-actin when not polymerized. Regulatory proteins and molecular motors constantly remodel the actin filaments and their dynamics have been studied in vivo [8], in reconstituted in vitro systems [2, 9], and in silico [10]. Actin filaments are capable of forming large-scale networks and can generate pushing, pulling, and resistive forces necessary for various cellular functions such as cell motility, mechanosensation, and tissue morphogenesis [8]. Therefore, insights into actin dynamics will advance our understanding of cellular physiology and associated pathological conditions [2, 11].

Actin filaments in cells are dynamic and strongly out of equilibrium. The filaments are semi-flexible, rod-like structures approximately $0.007\,\mu m$ in diameter and extending several microns in length, formed through the assembly of G-actin subunits [6, 7]. A filament has two ends, a barbed end and a pointed end, with distinct growth and decay properties. A filament length undergoes cycles of growth and decay fueled by an input of chemical energy, in the form of ATP, to bind and unbind actin monomers [6, 7]. The rates at which actin molecules bind and unbind from a filament have been measured experimentally [3, 12, 13]. The cell tightly regulates the number, density, length, and geometry of actin filaments [7]. In particular, the geometry of actin networks is controlled by a class of accessory proteins that bind to the filaments or their subunits. Through such interactions, accessory proteins are able to determine the assembly sites for new filaments, change the binding and unbinding rates, regulate the partitioning of polymer proteins between filaments and subunit forms, link filaments to one another, and generate mechanical forces [6, 7]. Actin filaments have been observed to organize into branched networks [14, 15], sliding bundles extending over long distances [16], and transient patterns including vortices and asters [17, 18].

To generate pushing forces for motility, the cell uses the energy of the growth or polymerization of F-actin [19, 20]. Actin polymerization powers the formation of flat cellular protrusions, known as lamellipodia, found at the leading edge of motile cells [8, 21]. Microscopy of the lamellipodial cytoskeleton has revealed multiple branched actin filaments [15]. The branching structure is governed by the Arp2/3 protein complex, which binds to an existing actin filament and initiates growth of a new "daughter" filament through a nucleation site at the side of preexisting filaments. Growth of the "daughter" filament occurs at a tightly regulated angle of 70° from the "parent" filament due to the structure of the Arp2/3 complex [22]. The directionality of pushing forces produced by actin polymerization originates from the uniform orientation of polymerizing actin filaments with their barbed ends towards the leading edge of the cell [8]. Here, cells exploit the polarity of filaments, since growth dynamics are faster at barbed ends than at pointed ends [23, 24]. Polymerization of individual actin filaments produces piconewton forces [25], with filaments organized into a branched network in lamellipodia or parallel bundles in filopodia [15]. The localized kinetics of growth, decay, and branching of a protrusive actin network provide the cell with the scaffold and the mechanical work needed for directed movement.

Many mathematical models have been developed to capture the structural formation and force generation of actin networks [20, 26, 27]. Due to the multiscale nature of actin dynamics, two main approaches are used: agent-based methods [27–29] and deterministic models using PDEs [30–33]. The effects of different molecular components (e.g., depolymerization, stabilizers) on the architecture of a protrusive actin network has been studied with detailed hybrid micro-macroscopic models [34, 35]. While both techniques are useful for understanding actin dynamics, each presents limitations. Agent-based models more closely capture the molecular dynamics of actin by explicitly considering the behavior of actin molecules through rules, such as, bind to the closest filament at a particular rate. In general, agent-based models simulate the spatiotemporal actions of certain microscopic entities, or "agents", in an effort to recreate and predict more complex large-scale behavior. In these simulations, agents behave autonomously and through simple rules prescribed at each time step. The technique is stochastic and can be interpreted as a coarsening of Brownian and Langevin dynamical models [36]. However, agent-based approaches are computationally expensive: at every time step, they specifically account for the movement and interaction of individual molecules, while also assessing the effects of spatial and environmental properties that ultimately result in the emergence of certain large-scale phenomena, such as crowding. Such approaches benefit from the direct relationship to experimental measurements of parameters, yet they present a further computational cost in that many instances of a simulation are needed for reliable statistical information. Agent or rule-based approaches have been used to reveal small-scale polymerization dynamics in actin polymer networks [26, 37], but due to the inherent computational complexity, it remains unclear how this information translates to higher length scales, such as the cell, tissue, or whole organism.

To overcome such computational costs and still gain a mechanistic understanding of actin processes, one approach is to write deterministic equations that "summarize" all detailed stochastic events. These approaches rely on differential equations to predict a coarse-grained biological behavior by assuming a well-mixed system where the molecules of interest exist in high numbers [20, 38–40] and the spread of the polymer network can be qualitatively approximated by a diffusion process [41, 42]. In continuum models, the stochastic behavior of the underlying molecules are typically ignored. While continuum models can be explored via traditional mathematical analysis, the challenge lies in determining the terms and parameters of these equations that are representative of the underlying physical system. Thus, these methods use phenomenological parameters of the actin network, such as bulk diffusion and reaction terms, that are less readily obtained experimentally. The relationship between molecular-level actin processes and cellular-level actin network behavior remains disconnected. This disconnect presents a unique challenge in modeling actin polymers in an active system across length scales.

In this work, we design a systematic and rigorous methodology to compare and connect actin molecular effects in agent-based stochastic simulations to macroscopic behavior in deterministic continuum equations. Measures from these distinct-scale models enable extrapolation from the molecular to the macroscopic scale by relating local actin dynamics to phenomenological bulk parameters. First, we characterize the dynamics of a protrusive actin network in free space using a minimal agent-based model for the branching of actin filaments from a single nucleation site based on experimentally measured kinetic rates. Second, in the macroscopic approach, we simulate the spread of actin filaments from a point source using a partial differential equation model. The model equation is derived from first principles of actin filament dynamics and is found to be Skellam's equation for unbounded growth of a species together with spatial diffusion. To compare the emergent networks, multiple instances of the agent-based approach are simulated, and averaged effective diffusion coefficient and reaction rate are extracted from the mean displacement of the advancing network front and from the averaged network density. We identify two concrete metrics, mean displacement and the averaged filament length density, that connect phenomenological bulk parameters in the reaction-diffusion systems to the molecular biochemical rates of actin binding, unbinding, and branching. Using sensitivity analysis on these measures, we demonstrate that the outward movement of the actin network is insensitive to changes in parameters associated with branching, while the bulk growth rate and diffusion coefficient do vary with changes in branching dynamics. We further find that the outward speed, growth rate constant, and effective diffusion increase with

F-actin polymerization rate but decrease with increasing depolymerization of actin filaments. By formalizing the relationship between micro- and macro-scale actin network dynamics, we demonstrate a nonlinear dependence of bulk parameters on molecular characteristics, indicating the need for careful model construction and justification when modeling the dynamics of actin networks.

2   Mathematical Models

2.1   Microscale Agent-Based Model

2.1.1   Model Description

We build a minimal, agent-based model sufficient to capture the local microstructure of a branching actin network. This model includes the dynamics of actin filament polymerization, depolymerization, and branching from a nucleation site [1, 2, 43, 44]. We treat F-actin filaments as rigid rods. Each actin filament has a base (pointed end) fixed in space and a tip (barbed end) capable of growing or shrinking due to the addition or removal of actin monomers, respectively. We assume that there is an unlimited pool of actin monomers available for filament growth, in line with normal, intracellular conditions [3]. For simplicity, we neglect the effects of barbed end capping, mechanical response of actin filaments, resistance of the plasma membrane, cytosolic flow, and molecular motors and regulatory proteins. Motivated by the short timescale of the initial burst of a growing actin network, we assume that the pointed end of actin filaments is stabilized at a nucleation site, and thus, do not account for the turnover dynamics at the pointed end. The physical setup of the model is similar to conditions associated with in vitro experiments, as well as initial actin network growth in cells, before components such as actin monomers become limiting.

2.1.2   Numerical Implementation

At the start of each simulation, an actin filament of length zero is assigned an angle of growth from the nucleation site (located at the origin) from a uniform random distribution. Once a filament is prescribed a direction of growth, it will not change throughout the time-evolution of that particular filament. At each subsequent time step in the simulation, there are four possible outcomes: (i) growth of the filament with probability p_{poly}, (ii) shrinkage of the filament with probability p_{depoly}, (iii) no change in filament length, or (iv) branching of a preexisting filament into a "daughter" filament with probability p_{branch} provided that the "parent" filament has reached a critical length L_{branch}, measured from the closest branch point. To determine which outcome occurs, two random numbers are independently generated for each filament. The first random number governs polymerization (i) or depolymerization (ii): if the random number is less than p_{poly}, polymerization occurs, and if greater than $1 - p_{depoly}$, depolymerization occurs. If the first random number is greater than or equal to p_{poly} and less than or equal to $1 - p_{depoly}$, then the filament neither polymerizes nor depolymerizes this time step, and therefore remains the same length (iii). Similarly, if the random number is simultaneously less than p_{poly} and greater than $1 - p_{depoly}$, then both polymerization and depolymerization occur within this time step, and therefore the filament remains the same length (iii). Both filament growth and shrinkage occur in discrete increments corresponding to the length of a G-actin monomer, $\Delta x = 0.0027\,\mu m$ [4]. We enforce that a filament of length zero cannot depolymerize.

The second random number pertains to filament branching (iv). For filaments of length greater than L_{branch}, a new filament can be initiated at a randomly oriented $70°$-angle from a preexisting filament tip in correspondence with the effect of Arp2/3 protein complex. If the second random number is less than the probability of branching, p_{branch}, for the given filament, then the filament will branch and create a "daughter" filament now capable of autonomous growth and branching. This branching potential models the biological effect of the Arp2/3 complex without explicitly including Arp2/3 concentration as a variable.

The step-wise process is repeated until the final simulation time is reached. Simulation steps are summarized graphically in Fig. 1. All parameters for the model are listed in Table 1. We calculate several different measurements from the microscale simulation, as described below.

2.1.3   Parameter Estimation

Actin dynamics have been extensively studied in vivo and in vitro, providing many rate constants used in this study. A $10\,\mu M$ actin monomer concentration elongates the barbed ends of F-actin filaments at a reported velocity of $0.3\,\mu m/s$ [3]. We use this measurement to calculate the polymerization probability, p_{poly}, via the formula:

$$\text{assembly rate} = \text{polymerization probability} \times \text{length added to filament}$$
$$\times \text{ number of timesteps per second} \qquad (1)$$

$$0.3\,\tfrac{\mu m}{s} = p_{poly} \times 0.0027\,\mu m \times \frac{1}{0.005\,s}\,, \qquad (2)$$

which implies that $p_{poly} = 0.56$. For simplicity, we round this probability to $p_{poly} = 0.6$ in the microscale model simulations. ADP-actin has a depolymerization rate of $4.0\,1/s$ at the barbed ends of actin filaments [3]. This measurement represents the rate of depolymerization of one actin subunit per second, thus a filament loses length at a rate of

$$4.0\,\tfrac{\text{subunit}}{s} \times 0.0027\,\tfrac{\mu m}{\text{subunit}} = 0.0108\,\tfrac{\mu m}{s}. \qquad (3)$$

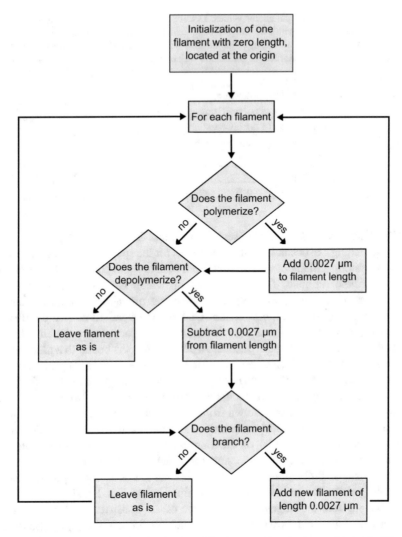

Fig. 1 Flow chart of the algorithm implemented for the agent-based microscale model. All steps following "Initialization" are repeated at every time step

To calculate the depolymerization probability, p_{depoly}, we use the analogous formula:

disassembly rate = depolymerization probability × length removed from filament

$$\times \text{ number of timesteps per sec} \tag{4}$$

$$0.0108 \, \tfrac{\mu m}{s} = p_{\text{depoly}} \times 0.0027 \, \mu m \times \frac{1}{0.005 \, s}, \tag{5}$$

which yields $p_{\text{depoly}} = 0.02$.

Table 1 Microscale model parameter values. Details on parameter estimation are available in Sect. 2.1.3. Values flagged with one star (*) were calculated from [3] and depend on the time step as indicated in Eqs. 2 and 5. The value flagged with a dagger (†) is motivated by literature measurements of actin filament length per branch which vary from 0.02 to 5 μm [15, 45–49]

Parameter	Meaning	Value
p_{branch}	Probability of branching	normal CDF
p_{poly}	Probability of polymerizing	0.6*
p_{depoly}	Probability of depolymerizing	0.02*
L_{branch}	Critical length before branching can occur	0.2 μm†
T_{end}	Total run time	10 s
Δt	Time step	0.005 s
n_{sim}	Number of independent simulations	10

Model parameter L_{branch} represents the critical length a filament must reach before branching can occur. Literature estimates for the spacing of branching Arp2/3 complexes along a filament vary widely, from 0.02 to 5 μm [15, 45–49]. We choose an intermediate estimate, $L_{branch} = 0.2$ μm, which is of similar order to the values from other studies [48, 49]. The branching probability, p_{branch}, is chosen from a cumulative distribution function (CDF) of the standard normal distribution with mean, $\mu = 2$ and standard deviation, $\sigma = 1$. For in vitro systems, branch formation is inefficient because once an Arp2/3 complex is bound to a filament, the reported branching rate is slow (estimated to be $0.0022 - 0.007$ s^{-1}) [49]. Given the relative dynamic scales of polymerization/depolymerization versus branching, we assume that (de)polymerization occurs at a prescribed rate, but because branching is infrequent, its probability is drawn from a distribution function.

The three microscale model probabilities are calculated in a time-step-dependent manner, such that the results of the microscale simulation are independent of the value of Δt, for a given time step for which calculated polymerization and depolymerization probabilities are not greater than 1. For example, polymerization and depolymerization probabilities are obtained using Eqs. 1 and 4. From Eqs. 1 and 2 we see that the largest possible value of Δt consistent with a probability less than or equal to 1 is 0.009 s. The branching probability, p_{branch}, is always obtained using the CDF described in the preceding paragraph, but branching is only allowed to happen at fixed time intervals of 0.005 s. Using the Δt value from Table 1, branching can occur every time step with probability p_{branch}. If, instead, $\Delta t = 0.0025$ s, branching can occur every other time step with probability p_{branch}. Simulations in this study were performed with a time step $\Delta t = 0.005$ to ensure that all results are internally consistent and comparable.

2.2 Macroscale Deterministic Model

2.2.1 Model Description

We model the growth and spread of a branching actin network through a reaction-diffusion form of the chemical species conservation equation derived in Sect. 2.2.2. This form is frequently known as Skellam's equation, applied to describe populations that grow exponentially and disperse randomly [50]. The two-dimensional Skellam's equation is

$$\frac{\partial \tilde{u}}{\partial t} = D \left(\frac{\partial^2 \tilde{u}}{\partial x^2} + \frac{\partial^2 \tilde{u}}{\partial y^2} \right) + r\tilde{u}. \tag{6}$$

In the context of our actin network, $\tilde{u}(x, y, t)$ is a dimensionless, normalized number density of polymerized actin monomers at location (x, y) and time t, D is the diffusion coefficient of the network (network spread), and r is the effective growth rate constant (network growth). More precise definitions of $\tilde{u}$ and r will be given in Sect. 2.2.2. Note that the diffusion coefficient is in reference to the bulk F-actin network spread, rather than representing Fickian behavior of monomers as has been done in previous literature [31, 40]. We use no flux boundary conditions in Eq. 6 to enforce no flow of actin across the cell membrane. For the initial condition of Eq. 6, we prescribe a point source at the origin.

2.2.2 Derivation of Reaction Term from First Principles

We present a derivation of the reaction term in Skellam's continuum description (Eq. 6) from simple kinetic considerations of actin filaments which include polymerization, depolymerization, and branching.

First, we write the molecular scheme for actin filament polymerization and depolymerization in the form of chemical equations. We denote a G-actin monomer in the cytoplasmic pool by M, an actin polymer chain consisting of $n - 1$ subunits by p_{n-1}, and a one monomer longer actin polymer chain by p_n. The process of binding and unbinding of an actin monomer is described by the following reversible chemical reaction:

$$M + p_{n-1} \underset{k_r}{\overset{k_f}{\rightleftharpoons}} p_n . \tag{7}$$

The constants k_f and k_r represent the forward and reverse rate constants, respectively, and encompass the dynamics that lead to the growth/shrinking of an actin filament.

Next, the biochemical reaction in Eq. 7 can be translated into a differential equation that describes rates of change of the F-actin network density. To write the corresponding equations, we first use the law of mass action which states that the rate of reaction is proportional to the product of the concentrations. Then, the rates of the forward (r_f) and reverse (r_r) reactions are:

$$r_f = k_f [M][p_{n-1}], \tag{8}$$

$$r_r = k_r [p_n], \tag{9}$$

where brackets denote concentrations of M, p_{n-1}, and p_n, in number per unit area. This single actin polymerization/depolymerization reaction can be extended to capture all actin filaments reacting simultaneously across the network as follows:

$$r_f = k_f [M][P_{n-1}], \tag{10}$$

$$r_r = k_r [P_n], \tag{11}$$

where we define

$$P_n = \sum_{i=2}^{n} i [p_i]. \tag{12}$$

Under the assumption that the forward and reverse reactions are each elementary steps, the net reaction rate is

$$r_{net} = r_f - r_r = k_f [M][P_{n-1}] - k_r [P_n]. \tag{13}$$

We note that the monomer concentration $[M]$ can be eliminated from Eq. 13 if it is expressed in terms of initial concentration of monomers in the cell cytoplasm $[M]_0$:

$$[M] = [M]_0 - \sum_{i=2}^{n} i [p_i] = [M]_0 - [P_n]. \tag{14}$$

Lastly, we note that $[P_n] = [P_{n-1}] + n [p_n]$. We can assume a minor contribution from actin polymers at this maximum length, such that $[P_n] \approx [P_{n-1}]$. Then, Eq. 13 becomes

$$r_{net} = k_f \left([M]_0 - [P_n]\right)[P_n] - k_r [P_n], \tag{15}$$

and can be further simplified if we divide both sides of the equation by $[M]_0$:

$$\frac{r_{net}}{[M]_0} = [M]_0 \frac{[P_n]}{[M]_0} \left[\left(1 - \frac{[P_n]}{[M]_0}\right) k_f - \frac{1}{[M]_0} k_r\right]. \tag{16}$$

We introduce variable $\tilde{u}$ as the polymerized actin concentration normalized by the initial monomeric actin concentration and r as an effective growth rate constant:

$$\tilde{u} = \frac{[P_n]}{[M]_0}, \quad r = k_f [M]_0. \tag{17}$$

The normalized net reaction rate in Eq. 16 simplifies to

$$\frac{r_{net}}{[M]_0} = r\tilde{u}(1 - \tilde{u}) - k_r\tilde{u} . \tag{18}$$

Describing the macroscale dynamics of polymerized actin concentration $[P_n]$ as simultaneously diffusing in two-dimensional space with diffusion coefficient D and undergoing molecular reactions with the net reaction rate r_{net} results in the following PDE:

$$\frac{\partial [P_n]}{\partial t} = D \left(\frac{\partial^2 [P_n]}{\partial x^2} + \frac{\partial^2 [P_n]}{\partial y^2} \right) + r_{net} . \tag{19}$$

The equation expressed in terms of variable $\tilde{u} = [P_n]/[M]_0$ becomes

$$\frac{\partial \tilde{u}}{\partial t} = D \left(\frac{\partial^2 \tilde{u}}{\partial x^2} + \frac{\partial^2 \tilde{u}}{\partial y^2} \right) + \frac{r_{net}}{[M]_0} . \tag{20}$$

Finally, substituting in the full form of the normalized net reaction rate from Eq. 18 produces

$$\frac{\partial \tilde{u}}{\partial t} = D \left(\frac{\partial^2 \tilde{u}}{\partial x^2} + \frac{\partial^2 \tilde{u}}{\partial y^2} \right) + r\tilde{u}(1 - \tilde{u}) - k_r\tilde{u} . \tag{21}$$

Special Cases
We consider two special cases of the kinetics derived above. In the case where monomeric actin concentration is much larger than polymerized actin concentration, the normalized net reaction rate in Eq. 18 becomes

$$\frac{r_{net}}{[M]_0} = r\tilde{u} - k_r\tilde{u} . \tag{22}$$

Further, based upon experimental measurements in [3] whereby the depolymerization rate is approximately three orders of magnitude slower than that of polymerization (i.e., slow reverse rate with $k_r \to 0$), the normalized net reaction rate in Eq. 18 simplifies to

$$\frac{r_{net}}{[M]_0} = r\tilde{u}(1 - \tilde{u}) . \tag{23}$$

Taking these two special cases together yields the following net reaction rate:

$$\frac{r_{net}}{[M]_0} = r\tilde{u} .$$

(24)

Substituting $r_{net}/[M]_0$ for the case of both unlimited monomers and slow depolymerization into Eq. 20 yields the same functional form as Skellam's equation for unbounded growth (Eq. 6):

$$\frac{\partial \tilde{u}}{\partial t} = D\left(\frac{\partial^2 \tilde{u}}{\partial x^2} + \frac{\partial^2 \tilde{u}}{\partial y^2}\right) + r\tilde{u} .$$

(25)

Conversely, substituting $r_{net}/[M]_0$ for only the case of slow depolymerization into Eq. 20 produces Fisher's equation for saturated growth:

$$\frac{\partial \tilde{u}}{\partial t} = D\left(\frac{\partial^2 \tilde{u}}{\partial x^2} + \frac{\partial^2 \tilde{u}}{\partial y^2}\right) + r\tilde{u}(1 - \tilde{u}) .$$

(26)

For the current system under study, the two aforementioned special case assumptions hold, in that the monomer pool is unlimited and the rate of polymerization far exceeds the rate of depolymerization. Therefore, the former equation (Skellam's) is chosen to model macroscale actin dynamics.

One of the main goals of this study is to compare the F-actin densities predicted by the microscale model and the macroscale model. We introduce the length density, $u(x, y, t)$, which is defined as the total length of actin filaments per unit area at the location (x, y) and time t. The length density is related to $\tilde{u}$ by

$$u = 0.0027 \,\mu\text{m} \times [M]_0 \times \tilde{u},$$

(27)

recalling from Sect. 2.1.2 that $0.0027 \,\mu\text{m}$ is the length added to an actin filament by a monomer. As $0.0027 \,\mu\text{m} \times [M]_0$ is a constant, u also satisfies Skellam's equation:

$$\frac{\partial u}{\partial t} = D\left(\frac{\partial^2 u}{\partial x^2} + \frac{\partial^2 u}{\partial y^2}\right) + ru.$$

(28)

We note that while $\tilde{u}$ is dimensionless, u is measured in length per unit area and thus, has units of $\mu\text{m}/\mu\text{m}^2 = 1/\mu\text{m}$. As it is more straightforward to calculate the length density of F-actin in the microscale model, we will use u instead of $\tilde{u}$.

2.2.3 Analytical Solution of PDE

The analytical solution to Eq. 28, given an initial point source at the origin is

$$u(x, y, t) = \frac{u_0}{4\pi Dt} \exp\left(rt - \frac{x^2 + y^2}{4Dt}\right),$$

(29)

where u_0 is the magnitude of the point source. For sufficiently large times t, the F-actin density in Eq. 29 propagates as a unidirectional wave moving at a constant speed v. To see this, we first fix u and solve for $x^2 + y^2$ from Eq. (29):

$$x^2 + y^2 = 4rDt^2 - 4Dt\left(\ln t + \ln\frac{4\pi uD}{u_0}\right). \tag{30}$$

Applying the limit of large time yields

$$\lim_{t\to+\infty}\frac{\sqrt{x^2+y^2}}{t} = \lim_{t\to+\infty} 2\sqrt{rD - \frac{D}{t}\left(\ln t + \ln\frac{4\pi uD}{u_0}\right)} = 2\sqrt{rD}. \tag{31}$$

For large enough t values, $u(x, y, t)$ is a traveling wave propagating with speed $v = 2\sqrt{rD}$. Indeed, in the stochastic simulations, we observe that the speed at which the periphery of the actin network advances is roughly constant (see Fig. 2c).

The wave-like behavior of an actin network has attracted considerable interest in recent years [51–54]. In [54], a variety of experimental and theoretical studies of actin traveling waves have been classified and reviewed. It is generally thought that actin waves result from the interplay between "activators" and "inhibitors" of actin dynamics modulated by regulatory proteins. Activation and inhibition are incorporated into our stochastic model by introducing the probabilities of branching, polymerization, and depolymerization. For a fixed set of parameter values, we can infer the values of r, D from the F-actin density averaged over many runs of the stochastic model. In addition, by varying these stochastic model parameters, we can gain insight into how they affect r, D, and the wave speed of the network.

3 Mathematical Methods

3.1 Measures to Connect Microscale Agent-Based and Macroscale Deterministic Models

In this section, we state the framework we developed to compare and connect the microscale agent-based approach in Sect. 2.1 to the macroscale continuum system in Sect. 2.2. From many instances of the stochastic simulation, the averaged mean displacement and network density are computed and used, together with the analytical solution of Skellam's equation in Eq. 29, to extract an effective bulk diffusion coefficient and unsaturated growth rate. These two quantities in the continuum model are completely identifiable by characteristics of the microscale dynamics.

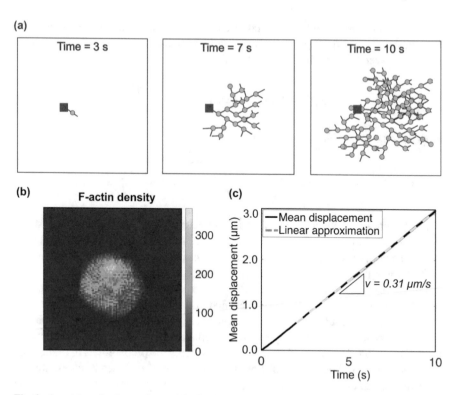

Fig. 2 Agent-based microscale model of a branching actin structure. (**a**) Resulting branching networks at different time instances of $t = 3, 7$, and 10 s. Light gray dots represent Arp2/3 protein complexes, dark gray squares indicate the initial nucleation site, and solid black lines denote F-actin filaments. (**b**) F-actin length density at $t = 10$ s. The filament length density is calculated from one realization of the stochastic system and measures the filament length per area. (**c**) Mean displacement of 10 independent realizations of the model (solid), with the corresponding best-fit linear approximation (dashed). The slope of the linear approximation corresponds to the wave speed of the leading edge of the network, $v = 0.31\,\mu$m/s. The parameters used in stochastic agent-based simulations are provided in Table 1. Mean displacement is discussed in more detail in Sect. 3.1.1

3.1.1 Mean Displacement

To track the movement of the actin network in the stochastic simulations, we define a "fictitious particle" to be the filament tip extending the greatest distance from the nucleation site at the origin. The position of this fictitious particle is calculated at each time step and we report the displacement of the fictitious particle as a function of time averaged over 10 independent realizations of the microscale algorithm (Fig. 2c). Note that the fictitious particle may not correspond to the same individual filament tip in consecutive time steps. We find the mean displacement over time is well-fitted by a linear function with a goodness-of-fit coefficient of $R^2 = 0.99996$; the linear correlation coefficient varied insignificantly with parameter variations. The slope of mean displacement of the fictitious particle in the stochastic simulations can be interpreted as the speed of propagation of the leading edge of the network. The speed of the network in the continuum approach was derived in Eq. 31. Combining these yields one link between the microscale simulations and the macroscale system:

$$v = \lim_{t \to \infty} \frac{x}{t} = 2\sqrt{rD} \, . \tag{32}$$

Here, v denotes the slope of the line which fits the mean displacement curve over time. We note that the important quantity for our analysis is "mean displacement", not mean-squared displacement, because our system does not undergo a purely diffusive process. Instead, we use mean displacement to extract the wave speed of an advancing network that undergoes both diffusion and density-dependent growth.

3.1.2 Derivation of r and D from the Network Density

To distinguish between the effective diffusion coefficient and growth rate in the wave speed in Eq. 32, a second measure is necessary to isolate the two parameters. To gain insight about what the second measure should be, we look to simplify the analytical solution of Skellam's equation to obtain an expression for one of the parameters. At an arbitrary time point t_i, and considering a cross-section of the solution (along $y = 0$ in Fig. 2b), Eq. 29 simplifies to

$$u(x, 0, t_i) = \frac{u_0}{4\pi D t_i} \exp\left(r t_i - \frac{x^2}{4D t_i}\right) . \tag{33}$$

Its first and second spatial derivatives are:

$$\frac{\partial}{\partial x}(u(x, 0, t_i)) = -\frac{u_0 x}{8\pi D^2 t_i^2} \exp\left(r t_i - \frac{x^2}{4D t_i}\right) , \tag{34}$$

$$\frac{\partial^2}{\partial x^2}(u(x, 0, t_i)) = -\frac{u_0}{8\pi D^3 t_i^3}\left(D t_i - \frac{x^2}{2}\right) \exp\left(r t_i - \frac{x^2}{4D t_i}\right) . \tag{35}$$

The spatial profile of the solution along a horizontal slice with $y = 0$ at $t_i = 10\,\text{s}$ together with its gradient are shown in Fig. 3a. At any point in time, there are three points of interest in the solution: the global maximum and the two inflection points. These points correspond to the zero of the gradient function (for the global maximum) and the global maxima and minima of the gradient function (for the inflection points). Critical point analysis shows that for Eq. 33 the global maximum of the solution occurs at $x = 0$, while the inflection points occur at $x = \pm\sqrt{2D t_i}$.

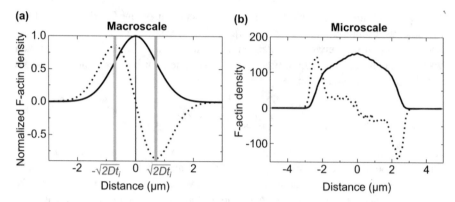

Fig. 3 Connection between macroscale deterministic and microscale agent-based models of a branching actin network. (**a**) Cross-section of the solution (solid line, Eq. 33) and first derivative (dotted line, Eq. 34) of the 2D solution to Skellam's equation with $y = 0$ at $t_i = 10\,\mathrm{s}$. The maximum of the solution (vertical black line) and the two inflection points (vertical gray lines) are indicated as three points of interest used to explicitly calculate the growth rate constant (r) from simulations of the microscale system. (**b**) F-actin length density averaged over 1000 independent runs of the microscale model (solid line), and its calculated gradient (dotted line)

To obtain an analytical expression for r, we find the global maximum of the solution curve at a time point t_1:

$$u(0, 0, t_1) = \frac{u_0}{4\pi D\, t_1}\, \exp(r\, t_1)\,. \tag{36}$$

A similar expression is obtained for t_2. Taking the ratio of $u(0, 0, t_1)$ and $u(0, 0, t_2)$, we conclude that

$$\frac{u(0, 0, t_1)}{u(0, 0, t_2)} = \frac{\frac{u_0}{4\pi D t_1}\, \exp(r\, t_1)}{\frac{u_0}{4\pi D t_2}\, \exp(r\, t_2)}\,, \tag{37}$$

$$= \frac{t_2}{t_1}\, \exp\left(r\,(t_1 - t_2)\right)\,, \tag{38}$$

$$\Rightarrow r = \frac{\ln\left(\frac{u(0,0,t_1)}{u(0,0,t_2)}\right) + \ln\left(\frac{t_1}{t_2}\right)}{t_1 - t_2}\,. \tag{39}$$

Here, $u(0, 0, t_1)$ and $u(0, 0, t_2)$ are the maximum values of our solution function at two different arbitrary time points. By averaging over many independent microscale simulations for a fixed set of parameters, we can approximate the maximum value of the actin network density. Thus, for two choices of time points t_1 and t_2, and the corresponding maximum values of actin network concentration averaged over many microscale simulations, we can explicitly calculate r, as follows:

$$r = \frac{\ln\left(\frac{\text{max value at } t_1}{\text{max value at } t_2}\right) + \ln\left(\frac{t_1}{t_2}\right)}{t_1 - t_2}\,. \tag{40}$$

Another method for estimating r is to use the inflection points, rather than the maximum value of the solution profile. The inflection points at time t_1 occur at $x = \pm\sqrt{2D\,t_1}$. At those points, the gradient of the solution curve is

$$\frac{\partial}{\partial x}(u(x,0,t_1))\Big|_{x=\pm\sqrt{2D t_1}} = -\frac{u_0}{8\pi D^2 t_1^2}\left(\pm\sqrt{2D t_1}\right)\exp\left(r\,t_1 - \frac{(\pm\sqrt{2D t_1})^2}{4D\,t_1}\right),$$

$$(41)$$

$$= \mp\frac{u_0}{4\sqrt{2\pi}(D\,t_1)^{3/2}}\exp\left(r\,t_1 - \frac{1}{2}\right), \tag{42}$$

$$= \mp\left(\frac{u_0\,e^{-1/2}}{4\sqrt{2\pi}}\right)\frac{\exp(r\,t_1)}{(D\,t_1)^{3/2}}. \tag{43}$$

Note the first expression on the right-hand side is constant, while the second expression depends on the parameters as well as the choice of a time point. We have a similar expression at time point t_2. Reminding ourselves that $\frac{\partial}{\partial x}(u(x,0,t_i))\Big|_{x=\pm\sqrt{2Dt_i}}$ is simply the maximum gradient of the actin network at time t_i, we take the ratio of the expressions at times t_1 and t_2:

$$\frac{\text{max gradient at } t_1}{\text{max gradient at } t_2} = \frac{\mp\left(\frac{u_0\,e^{-1/2}}{4\sqrt{2\pi}}\right)\frac{\exp(r\,t_1)}{(D\,t_1)^{3/2}}}{\mp\left(\frac{u_0\,e^{-1/2}}{4\sqrt{2\pi}}\right)\frac{\exp(r\,t_2)}{(D\,t_2)^{3/2}}}, \tag{44}$$

$$= \exp\left(r\,(t_1 - t_2)\right)\left(\frac{t_2}{t_1}\right)^{3/2}, \tag{45}$$

$$\Rightarrow r = \frac{\ln\left(\frac{\text{max gradient at } t_1}{\text{max gradient at } t_2}\right) + \frac{3}{2}\ln\left(\frac{t_1}{t_2}\right)}{t_1 - t_2} \tag{46}$$

This provides a second method to explicitly calculate r from stochastic runs of the microscale model.

To extract the growth rate constant, r, from the microscale agent-based system using either Eq. 40 or 46 entails the calculation of the length density u at a time point t_i, and we now address how this is done. We consider 1000 independent realizations of the microscale model. The computational domain $[-5, 5\,\mu\text{m}] \times [-5, 5\,\mu\text{m}]$ is uniformly subdivided into 100×100 boxes of size $0.1 \times 0.1\,\mu\text{m}$. In each run of the microscale model and for every discretized box, we calculate the length density at the center of the box at time t_i. To be more precise, we consider the example in Fig. 4, where a $0.1 \times 0.1\,\mu\text{m}$ grid box centered at (x, y) at time t_i in one simulation of the microscale model is shown. There are three filaments that "cross" the box, that is, a portion of each of the three filaments is contained in the box. We calculate the length of each portion, add up the three lengths, and set the length density at (x, y) and t_i in the current run of the microscale model to be the sum divided by 0.01,

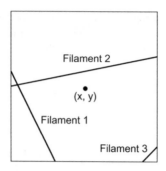

Fig. 4 A $0.1 \times 0.1 \,\mu$m grid box of an instance of the microscale stochastic model. The grid box is centered at (x, y) with three filaments partially contained in it

the area of the box. To obtain $u(x, y, t_i)$, we average the densities at (x, y) and t_i calculated in all 1000 independent simulations. The solid curve in Fig. 3b represents the averaged F-actin density at $(x, 0)$ and $t_i = 10$ s with default parameters provided in Table 1. To compute its spatial gradient, represented by the dashed line in Fig. 3b, we use centered differences.

Once we obtain r, and using the slope of the mean displacement of the network front, we can isolate the effective diffusion coefficient from Eq. 32 as

$$D = \frac{v^2}{4r} .$$ (47)

3.2 Sensitivity Analysis

To determine how microscale rates affect macroscale network behavior, we perform a series of sensitivity analyses. Specifically, we focus on three macroscopic measures: the wave speed of the advancing actin front (v), the effective growth rate constant (r), and the diffusion coefficient of the actin network (D) (Fig. 5). The microscale parameters varied are the critical length required for a filament to branch (L_{branch}), the polymerization probability (p_{poly}), the depolymerization probability (p_{depoly}), and the mean (μ) and standard deviation (σ) of the cumulative distribution function for filament branching probability (p_{branch}). As several of these parameters simultaneously influence the architecture of the network, three groups of parameters are established for analysis: L_{branch}, μ vs. σ, and p_{poly} vs. p_{depoly}. For each set of parameter runs, all other parameters are fixed at their default values in Table 1. Parameters are varied over the following ranges: $0.15 \leq L_{branch} \leq 1.2 \,\mu m$, $0 \leq p_{poly} \leq 0.75, 0 \leq p_{depoly} \leq 1, 0 \leq \mu \leq 5$, and $0 \leq \sigma \leq 5$.

The lower bound for the critical branching length is chosen to ensure computational tractability – as this value is lowered further, the network density continues to grow exponentially and becomes computationally demanding. The upper bound

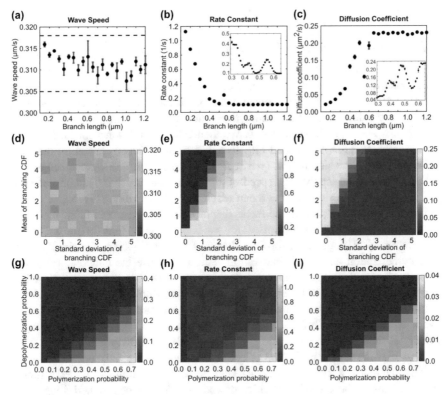

Fig. 5 Sensitivity analysis of the wave speed (v, μm/s), network growth rate (r, 1/s), and diffusion coefficient (D, μm²/s) as microscale parameters are varied. Effect of critical branching length (L_{branch}) on extracted (**a**) wave speed, (**b**) rate constant, and (**c**) diffusion coefficient. Black dots indicate the mean of 3 independent runs and red bars indicate standard error. The dotted lines serve as visual aids for the range of values for resulting wave speeds. The insets in (**b**) and (**c**) are more refined parameter variations for critical branching lengths between 0.3 and 0.7 μm. The horizontal axes of the insets represent critical branching lengths, while the vertical axes are the growth rate constant and diffusion coefficient, respectively. Effect of parameters associated with the branching probability with mean (μ) and standard deviation (σ) on calculated (**d**) wave speed, (**e**) network growth rate, and (**f**) diffusion coefficient. Effect of polymerization and depolymerization probabilities, (p_{poly} and p_{depoly}, respectively), on extracted (**g**) wave speed, (**h**) network growth rate, and (**i**) diffusion coefficient

for branching length is selected to capture the leveling-off behavior in Fig. 5b, c. Further, the interval for critical branching length encompasses many of the experimentally measured lengths. The upper bound for p_{poly} is again chosen to ensure computational tractability – a high polymerization rate with simultaneous low depolymerization rate increases the computational cost. Lastly, the two parameters associated with the cumulative distribution function for branching probability are non-negative. Their upper bound is arbitrary, yet importantly captures the essential trends in the network behavior in Fig. 5e, f.

4 Results

4.1 Micro-to-Macroscale Connection

To connect the dynamics of a branched actin network across the distinct scales, we simulate 1000 runs of the microscale agent-based model and record the resulting average actin density over a computational domain $[-5, 5\,\mu m] \times [-5, 5\,\mu m]$ at time points $t = 7, 8, 9, 10$, and $15\,s$. We then calculate the wave speed, followed by the network growth rate and effective diffusion coefficient based on the data collected from these simulations (see Sect. 3.1). Specifically, the wave speed results are calculated from the average speed of a fictitious particle at the leading edge of the network, or the slope of the mean displacement over time (Fig. 5a, d, g). To obtain the slope, the plot of the mean displacement over time is fitted by a line with a goodness-of-fit coefficient of $R^2 = 0.9999 - 0.99999$ for most choices of parameters. The goodness-of-fit is lower, $R^2 = 0.5 - 0.6$, for similar rates of polymerization and depolymerization. This is because the network undergoes periods of growth followed by decay and the effect is even more dramatic when depolymerization rate is faster than polymerization rate. In this case we report that the wave speed is zero since overall the network cannot grow. The maximum averaged filament length density or the maximum rate of change of the averaged density at time points $t_1 = 9$ and $t_2 = 10\,s$ yields the growth rate constant from Eq. 40 (Fig. 5b, e, h). Lastly, the effective diffusion of the bulk network can be readily calculated according to Eq. 47 (Fig. 5c, f, i). We simulate the PDE model in Eq. 28 using the diffusion coefficient and growth rate determined above, and compare the density predicted by the continuum model to the averaged density produced by the agent-based model (Fig. 6). We choose $u_0 = 0.1$ in Eq. 29, which seems to give the best overall fit to the microscale model.

We report that the wave speed is $v \approx 0.31\,\mu m/s$, while the rate constant is $r \approx 0.88\,1/s$ using Eq. 40, and the diffusion constant is $D \approx 0.03\,\mu m^2/s$. The averaged actin densities along a cross-section with $y = 0$ at times $t = 7, 8, 9, 10, 15\,s$ produced by the agent-based simulations are plotted in light gray in Fig. 6, while the corresponding densities from Skellam's equation are shown in dark gray.

4.2 Results of the Sensitivity Analysis

We find that the wave speed of the actin network is largely unaffected by branching parameters, either the critical branching length or the mean and standard deviation of the cumulative distribution function for branching probability (Fig. 5a, d). However, the wave speed does depend on polymerization and depolymerization probabilities, which dictate the rate of growth of filaments. This result is reasonable given our initial assumption of unlimited resources – branching events control the spatial distribution of the network, while the rates of filament length change dictate

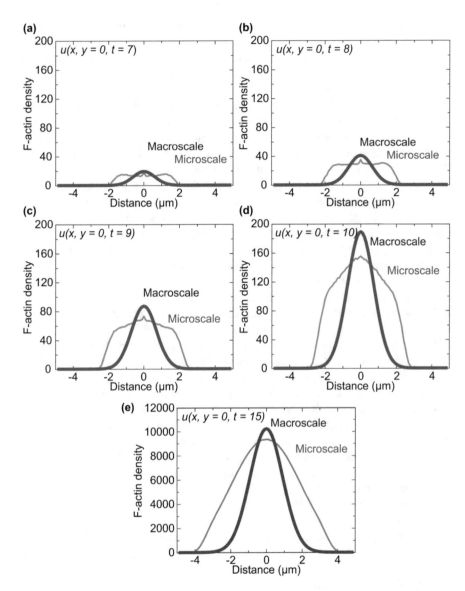

Fig. 6 Comparison of F-actin density profiles obtained from agent-based simulations (light) and analytical solution to Skellam's equation (dark) along a cross-section with $y = 0$ and time instances of (**a**) 7 s, (**b**) 8 s, (**c**) 9 s, (**d**) 10 s, and (**e**) 15 s. Macroscale parameters for the solution to Skellam's equation, $D = 0.03 \, \mu\text{m}^2/\text{s}$ and $r = 0.88 \, 1/\text{s}$, were calculated from averaged agent-based simulations as described in Sect. 3

the speed of network extension. Thus, increasing the polymerization probability increases the rate at which the network grows outwardly, i.e., the wave speed (Fig. 5g). Similarly, as the depolymerization probability increases, actin filaments are less likely to grow until eventually the overall growth of the network is arrested (top left corner on Fig. 5g).

In contrast, the effective growth rate constant and diffusion coefficient are affected by all five microscale kinetic parameters while keeping the wave speed of the network constant (Fig. 5, last two columns). We find that the growth rate constant is approximately inversely proportional to the critical branching length (Fig. 5b). The physical intuition is that the growth rate constant is a measure of the number of tips available for growth events (Eq. 7). For large critical branching lengths, the network is composed of a small number of filaments that persistently grow until the length condition for a branching event is met. Since a branching event can only occur at tips of filaments in our model, for large critical lengths only a small number of tips are available as branching sites, resulting in a small number of sites undergoing growth. The growth rate ranges between roughly 0.1 and 1.2 1/s, where the lower bound is attained for critical branching lengths over $\sim$0.5 μm. Variation of the two parameters of the branching cumulative distribution function – mean and standard deviation – produce a transition from the lower to the upper bound of growth rate (Fig. 5e). For a fixed but low standard deviation of the cumulative distribution function, decreasing the mean shifts the cumulative distribution to the left and thus increases the probability that a branching event can occur. However, for a fixed mean, increasing the standard deviation of the cumulative distribution decreases its slope, and thus results in larger probability for a filament to branch. Taken together, a left-shifted, shallow branching cumulative distribution function results in more branching events, and thus more filament tips that can undergo growth (bottom right in Fig. 5e). A similar but more gradual transition is found with changes in polymerization and depolymerization rates (Fig. 5h). The upper bound of the effective growth rate occurs for rapidly growing actin filaments, where polymerization probability is high but depolymerization probability is low (Fig. 5h, bottom right). This is due to filaments with a high polymerization probability reaching their critical lengths more quickly, allowing branching to start sooner, and the network to spread out more quickly. To summarize, our findings indicate that the growth rate of the leading edge of the network is dependent on the growth and decay rates of the filaments, but also on the number of filament tips available for binding of G-actin monomers.

The effective diffusion constant is calculated from the wave speed and growth rate constant using Eq. 47. Thus, to maintain a constant wave speed as critical branching length and branching probability are varied (Fig. 5a, d), the parameter dependence of the diffusion constant must complement the parameter dependence of the rate constant (Fig. 5b–c, e–f). We find that the diffusion coefficient increases with increasing critical branching length until it reaches a plateau value of approximately 0.25 μm^2/s for branching lengths over $\sim$0.5 μm (Fig. 5c). In this regime, the network is composed of few, long filaments that grow persistently since branching does not occur until a large critical length is reached. The network front moves in an approximately ballistic way rather than a diffusive, space-exploring way. We note that the transition to a plateau occurs at a similar critical branching length of 0.5 μm for both the diffusion constant and growth rate constant because the wave speed is constant at this critical branching length (in fact, it is constant across all branching length values). A sharp transition in the diffusion constant is reported as the branching probability parameters – mean and standard deviation – are varied (Fig. 5f). A high mean coupled with a low standard deviation results in a cumulative distribution function that is steep and shifted to the right. This results in a lower probability to branch, which means that individual filaments grow more persistently, and the effective diffusion of the actin network from the nucleation site is faster (top left in Fig. 5f). Reducing the mean or increasing the standard deviation increases the probability to branch, which results in a denser actin network that does not diffuse as far from the nucleation site (see smaller diffusion coefficients in bottom and right of Fig. 5f). For fixed branching parameters, the effective diffusion can be slightly increased through faster growth of the filament, or slightly decreased with faster decay of the filament length (Fig. 5i). Specifically, permissible diffusion constants ranges between 0.005 and 0.035 $\mu m^2/s$.

5   Discussion

Distinct F-actin density profiles arise from the stochastic, microscale simulations and the deterministic, macroscale model (Fig. 6). In the microscale approach, the flat filamentous actin profile with sharp shoulders at the boundary indicates that the outward drive of the advancing actin network dominates over the filament production term. In contrast, the continuum model reveals a more balanced outward diffusion with reaction production at the origin, as evidenced by the smooth profile growing in both radial extent and magnitude. This functional mismatch between the microscale and macroscale results could be due to assumptions of either model.

In the microscale model, we have only incorporated polymerization and depolymerization dynamics from the barbed actin filament end, with Arp2/3-mediated branching, while neglecting molecular motors and regulatory proteins involved in actin dynamics. We have also neglected the effects of barbed end capping and mechanical properties of actin filaments. Future extensions of the model will incorporate a wider set of proteins acting on the actin filaments, as well as the effect of these proteins on the network structure and outgrowth. Moreover, we aim to investigate how limited availability, or a small finite pool, of G-actin monomers and Arp2/3 complexes affects the resulting network architecture. This resource-limited case presents a study relevant for various biological conditions.

With the macroscale model, we have arrived at the simplified reaction-diffusion equation in Eq. 6 by incorporating two main assumptions on the reaction term: (1) initial monomeric actin concentration far exceeds that of polymerized actin and (2) slow reverse reaction (depolymerization) rate at the barbed end. Assumption (1) is mathematically expressed by removing the saturating effect of the forward reaction term (polymerization). Physically, this implies an unlimited pool of actin monomers is available to be polymerized into actin filaments. Qualitatively, limiting monomeric actin would result in slower network growth but potentially similar resultant network morphology. However, this may invalidate the second assumption, as the forward rate becomes comparable in magnitude to the reverse rate. In assumption (2), we rely on experimental measurements of actin polymerization and depolymerization rates [3]. The more significant role of polymerization is recapitulated in our calculated wave speed from the microscale simulation, which is $0.31\,\mu$m/s, in close accord with the empirical polymerization rate of $0.3\,\mu$m/s reported in [3]. The combined effect of implementing assumptions (1) and (2) is a higher net reaction rate, or a higher production of polymerized actin. Without incorporating these assumptions, the peak in polymerized actin at the domain center (Fig. 6) may decrease, resulting in shoulders that more closely resemble those from the microscale model. However, the hurdle to incorporating the full reaction term (Eq. 18) is that there is no analytical solution for Eq. 21, and thus no direct connection to the microscale model outputs. Specifically, the value of effective rate constant r is calculated by Eqs. 40 and 46, and the traveling wave nature of Skellam's equation at long times provides the wave speed v (Eq. 32), from which diffusion coefficient D can be directly calculated (Eq. 47). This leads to the main assumption of the macroscale model, whereby the form of the PDE (Eq. 6) assumes that actin network dynamics can be captured by a combination of reactive and diffusive components. However, the absence of the shoulders in the density profile present in the microscale model (Fig. 6) suggests that the PDE form can be modified for improved matching across the two scales. The absence of these shoulders may be an artifact of this PDE form. Thus, other PDE forms are being pursued for future work. Finally, a spatially-dependent reaction term may be incorporated to correct for the fact that the polymerization/depolymerization and especially branching reactions are not truly homogeneous reactions occurring throughout the bulk phase of the system, but rather, there is a distinct location dependence as to where the reaction is taking place (e.g., only at the filament tips or with a minimum spacing).

Taken together, these results suggest that great care must be taken to ensure models of actin dynamics are consistent with the underlying physical system. Here, we propose a methodology to compare microscale stochastic approaches to macroscale PDE models in order to directly correlate kinetic rates like binding and unbinding rates to macroscopic parameters like diffusion and saturated growth coefficients. These concrete metrics will connect phenomological diffusion coefficient and reaction constants in PDEs to experimentally measurable molecular rates.

Acknowledgments The work described herein was initiated during the Collaborative Workshop for Women in Mathematical Biology hosted by the Institute for Pure and Applied Mathematics at the University of California, Los Angeles in June 2019. Funding for the workshop was provided by IPAM, the Association for Women in Mathematics' NSF ADVANCE "Career Advancement for Women Through Research-Focused Networks" (NSF-HRD 1500481) and the Society for Industrial and Applied Mathematics. The authors thank the organizers of the IPAM-WBIO workshop (Rebecca Segal, Blerta Shtylla, and Suzanne Sindi) for facilitating this research.

R.L.P. is supported by the NSF Graduate Research Fellowships (NSF DGE 1752814). M.W.R. is supported in part by NSF DMS-1818833. A.T.D. is supported by NSF DMS-1554896.

References

1. U.S. Schwarz and M.L. Gardel. United we stand – integrating the actin cytoskeleton and cell–matrix adhesions in cellular mechanotransduction. *J. Cell Sci.*, 125:3051–3060, 2012.
2. J. Stricker, T. Falzone, and M. Gardel. Mechanics of the F-actin cytoskeleton. *J. Biomech.*, 43:9, 2010.
3. T.D. Pollard and G.G. Borisy. Cellular motility driven by assembly and disassembly of actin filaments. *Cell*, 112:453–465, 2003.
4. H. Lodish, A. Berk, S.L. Zipursky, et al. *Molecular Cell Biology, 4th ed.* W. H. Freeman, New York, USA, 2000.
5. B. Alberts, A. Johnson, J. Lewis, D. Morgan, M. Raff, K. Roberts, and P. Walter, editors. *Intracellular membrane traffic*. Garland Science, New York, 2008.
6. D. Bray. *Cell Movements: From molecules to motility, 2nd ed.* Garland Science, New York, USA, 2001.
7. J. Howard. *Mechanics of motor proteins and the cytoskeleton*. Sinauer, Sunderland, USA, 2001.
8. T. Svitkina. The actin cytoskeleton and actin-based motility. *Cold Spring Harb. Perspect. Biol.*, 10:a018267, 2018.
9. M.H. Jensen, E.J. Morris, and D.A. Weitz. Mechanics and dynamics of reconstituted cytoskeletal systems. *Biochim. Biophys. Acta*, 1853:3038–3042, 2015.
10. D. Holz and D. Vavylonis. Building a dendritic actin filament network branch by branch: models of filament orientation pattern and force generation in lamellipodia. *Biophys. Rev.*, 10:1577–1585, 2018.
11. M. Kavallaris. *Cytoskeleton and Human Disease*. Humana Press, Totowa, USA, 2012.
12. L. Blanchoin, R. Boujemaa-Paterski, C. Sykes, and J. Plastino. Actin dynamics, architecture, and mechanics in cell motility. *Physiol. Rev.*, 94:235–263, 2014.
13. J.A. Cooper T.D. Pollard. Actin and actin-binding proteins. a critical evaluation of mechanisms and functions. *Annu. Rev. Biochem.*, 55:987–1035, 1986.
14. R.D. Mullins, J.A. Heuser, and T.D. Pollard. The interaction of Arp2/3 complex with actin: nucleation, high affinity pointed end capping, and formation of branching networks of filaments. *Proc. Natl. Acad. Sci. USA*, 95:6181–6186, 1998.
15. T.M. Svitkina and G.G. Borisy. Arp2/3 complex and actin depolymerizing factor/cofilin in dendritic organization and treadmilling of actin filament array in lamellipodia. *J. Cell Biol.*, 145:1009–1026, 1999.
16. R. Cooke. The sliding filament model 1972-2004. *J. Gen. Physiol.*, 123:643–656, 2004.
17. F.J. Nédélec, T. Surrey, A.C. Maggs, and S. Leibler. Self-organization of microtubules and motors. *Nature*, 389:305–308, 1997.
18. T. Surrey, F. Nédélec, S. Leibler, and E. Karsenti. Properties determining self-organization of motors and microtubules. *Science*, 292:1167–1171, 2001.
19. T.D. Pollard. Actin and actin-binding proteins. *Cold Spring Harb. Perspect. Biol.*, 8:1–17, 2016.
20. A. Mogilner and G. Oster. Cell motility driven by actin polymerization. *Biophys J.*, 71:3030–3045, 1996.
21. M. Abercrombie, J.E. Heaysman, and S.M. Pegrum. The locomotion of fibroblasts in culture ii. 'ruffling". *Exp Cell Res*, 60:437–444, 1970.
22. E.D. Goley and M.D. Welch. The Arp2/3 complex: an actin nucleator comes of age. *Nat. Rev. Mol. Cell Biol.*, 7:713–726, 2006.

23. H.E. Huxley. Electron microscope studies on the structure of natural and synthetic protein filaments from striated muscle. *J. Mol. Biol.*, 7:281–308, 1963.

24. D.T. Woodrum, S.A. Rich, and T.D. Pollard. Evidence for biased bidirectional polymerization of actin filaments using heavy meromyosin prepared by an improved method. *J. Cell Biol.*, 67:231–237, 1975.

25. T.D. Pollard D.R. Kovar. Insertional assembly of actin filament barbed ends in association with formins produces piconewton forces. *Proc Natl Acad Sci*, 101:14725–14730, 2004.

26. X. Wang and A.E. Carlsson. A master equation approach to actin polymerization applied to endocytosis in yeast. *PLOS Comput. Biol.*, 13:e1005901, 2017.

27. K. Popov, J. Komianos, and G.A. Papoian. MEDYAN: Mechanochemical simulations of contraction and polarity alignment in actomyosin networks. *PLOS Comput. Biol.*, 12:e1004877, 2016.

28. V. Wollrab, J.M. Belmonte, L. Baldauf, M. Leptin, F. Nédélec, and G.H. Koenderink. Polarity sorting drives remodeling of actin-myosin networks. *J Cell Sci.*, 132:jcs219717, 2018.

29. S. Dmitrieff and F. Nédélec. Amplification of actin polymerization forces. *J Cell Biol.*, 212:763–766, 2016.

30. L. Edelstein-Keshet and G.B. Ermentrout. A model for actin-filament length distribution in a lamellipod. *J. Math. Biol*, 43:325–355, 2001.

31. A. Mogilner and L. Edelstein-Keshet. Regulation of actin dynamics in rapidly moving cells: A quantitative analysis. *Biophys. J.*, 83:1237–1258, 2002.

32. A. Gopinathan, K.-C. Lee, J.M. Schwarz, and A.J. Liu. Branching, capping, and severing in dynamic actin structures. *Phys. Rev. Lett.*, 99:058103, 2007.

33. T. Kim, W. Hwang, H. Lee, and R.D. Kamm. Computational analysis of viscoelastic properties of crosslinked actin networks. *PLOS Comput. Biol.*, 5:e1000439, 2009.

34. F. Huber, J. Käs, and B. Stuhrmann. Growing actin networks form lamellipodium and lamellum by self-assembly. *Biophys. J.*, 95:5508–5523, 2008.

35. B. Stuhrmann, F. Huber, and J. Käs. Robust organizational principles of protrusive biopolymer networks in migrating living cells. *PLoS ONE*, 6:e14471, 2011.

36. J. Jeon, N.R. Alexander, A.M. Weaver, and P.T. Cummings. Protrusion of a virtual model lamellipodium by actin polymerization: A coarse-grained langevin dynamics model. *J. Stat. Phys.*, 133:79–100, 2008.

37. J. Weichsel and U.S. Schwarz. Two competing orientation patterns explain experimentally observed anomalies in growing actin networks. *Proc. Natl. Acad. Sci. USA*, 107:6304–6309, 2010.

38. M. Malik-Garbi, N. Ierushalmi, S. Jansen, E. Abu-Shah, B.L. Goode, A. Mogilner, and K. Keren. Scaling behaviour in steady-state contracting actomyosin networks. *Nat. Phys.*, 15:509–516, 2019.

39. E.A. Vitriol, L.M. McMillen, M. Kapustina, S.M. Gomez, D. Vavylonis, and J.Q. Zheng. Two functionally distinct sources of actin monomers supply the leading edge of lamellipodia. *Cell Rep.*, 11:433–445, 2015.

40. I.L. Novak, B.M. Slepchenko, and A. Mogilner. Quantitative analysis of G-actin transport in motile cells. *Biophys. J.*, 95:1627–1638, 2008.

41. E.L. Barnhart, K-C Kun-Chun Lee, K. Keren, A. Mogilner, and J.A. Theriot. An adhesion-dependent switch between mechanisms that determine motile cell shape. *PLoS Biol.*, 9:e1001059, 2011.

42. A. Buttenschön and L. Edelstein-Keshet. Correlated random walks inside a cell: actin branching andmicrotubule dynamics. *J. of Math. Biol.*, 79:1953–1972, 2019.

43. K. Rottner, J. Faix, S. Bogdan, S. Linder, and E. Kerkhoff. Actin assembly mechanisms at a glance. *J. Cell Sci.*, 130:3427–3435, 2017.

44. J.A. Theriot, J. Rosenblatt, D.A. Portnoy, P.J. Goldschmidt-Clermont, and T.J. Mitchison. Involvement of profilin in the actin-based motility of L. monocytogenes in cells and in cell-free extracts. *Cell*, 76:505–517, 1994.

45. A. Mogilner. Mathematics of cell motility: have we got its number? *J. Math. Biol.*, 58:105–134, 2009.

46. K.J. Amann and T.D. Pollard. Direct real-time observation of actin filament branching mediated by Arp2/3 complex using total internal reflection fluorescence microscopy. *Proc. Natl. Acad. Sci. USA*, 98:15009–15013, 2001.

47. M.H. Jensen, E.J. Morris, R. Huang, G. Rebowski, R. Dominguez, D.A. Weitz, J.R. Moore, and C-L.A. Wang. The conformational state of actin filaments regulates branching by actin-related protein 2/3 (Arp2/3) complex. *J. Biol. Chem.*, 287:31447–31453, 2012.

48. M. Vinzenz, M. Nemethova, F. Schur, J. Mueller, A. Narita, E. Urban, C. Winkler, C. Schmeiser, S.A. Koestler, K. Rottner, G.P. Resch, Y. Maeda, and J.V. Small. Actin branching in the initiation and maintenance of lamellipodia. *J. Cell Sci.*, 125:2775–2785, 2012.

49. B.A. Smith, K. Daugherty-Clarke, B.L. Goode, and J. Gelles. Pathway of actin filament branch formation by Arp2/3 complex revealed by single-molecule imaging. *Proc. Natl. Acad. Sci. USA*, 110:1285–1290, 2013.

50. J.G. Skellam. Random dispersal in theoretical populations. *Biometrika*, 38:196–218, 1951.

51. M.G. Vicker. Eukaryotic cell locomotion depends on the propagation of self-organized reaction–diffusion waves and oscillations of actin filament assembly. *Exp. Cell Res.*, 275:54–66, 2002.

52. T. Bretschneider, K. Anderson, M. Ecke, A. Müller-Taubenberger, B. Schroth-Diez, H.C. Ishikawa-Ankerhold, and G. Gerisch. The three-dimensional dynamics of actin waves. *Biophys. J.*, 96:2888–2900, 2009.

53. A.E. Carlsson. Dendritic actin filament nucleation causes traveling waves and patches. *Phys. Rev. Lett.*, 104:228102, 2010.

54. J. Allard and A. Mogilner. Traveling waves in actin dynamics and cell motility. *Curr. Opin. Cell Biol.*, 25:107–115, 2013.

Modeling RNA:DNA Hybrids with Formal Grammars

Nataša Jonoska, Nida Obatake, Svetlana Poznanović, Candice Price, Manda Riehl, and Mariel Vazquez

Abstract R-loops are nucleic acid structures consisting of a DNA:RNA hybrid and a DNA single strand. They form naturally during transcription when the nascent RNA hybridizes to the template DNA, forcing the coding DNA strand to wrap around the RNA:DNA duplex. Although formation of R-loops can have deleterious effects on genome integrity, there is evidence of their role as potential regulators of gene expression and DNA repair. Here we initiate an abstract model based on formal grammars to describe RNA:DNA interactions and the formation of R-loops. Separately we use a sliding window approach that accounts for properties of the DNA nucleotide sequence, such as C-richness and CG-skew, to identify segments favoring R-loops. We evaluate these properties on two DNA plasmids that are known to form R-loops and compare results with a recent energetics model from the Chédin Lab. Our abstract approach for R-loops is an initial step toward a more sophisticated

N. Jonoska (✉)
University of South Florida, Tampa, FL, USA
e-mail: jonoska@math.usf.edu

N. Obatake
Texas A&M University, College Station, TX, USA
e-mail: nida@math.tamu.edu

S. Poznanović
Clemson University, Clemson, SC, USA
e-mail: spoznan@clemson.edu

C. Price
Smith College, Northampton, MA, USA
e-mail: cprice@smith.edu

M. Riehl
Rose-Hulman Institute of Technology, Terre Haute, IN, USA
e-mail: riehl@rose-hulman.edu

M. Vazquez
University of California, Davis, CA, USA
e-mail: mariel@math.ucdavis.edu

© The Association for Women in Mathematics and the Author(s) 2021
R. Segal et al. (eds.), *Using Mathematics to Understand Biological Complexity*,
Association for Women in Mathematics Series 22,
https://doi.org/10.1007/978-3-030-57129-0_3

framework which can take into account the effect of DNA topology on R-loop formation.

1 Introduction

RNA can have significant regulatory roles in biological processes such as gene expression, gene inhibition and others (reviewed in the special issue [19]). Recently, some interest has shifted towards the regulatory role of the transcript RNA, often assumed to be just an intermediate towards the protein coding mRNA. In particular, formation of R-loops is seen as a major factor in the RNA transcript involvement in DNA repair [23].

R-loops are three-stranded hybrid structures consisting of an RNA:DNA duplex, and a displaced single strand of DNA (illustrated in Fig. 1). Experimental results indicate the prevalence of R-loops in vastly different genomes. In particular R-loops have been shown to occur with surprising regularity at highly conserved hotspots throughout mammalian genomes [2]. A high throughput sequencing method that can provide genome-wide profiling of R-loops showed that up to 5% of the human genome has the potential of forming R-loops [18]. While R-loops seem to be the most abundant non-B DNA structures found to date (reviewed in [2]), little is known about their function, their mechanism of formation, or their geometry and topology. Most R-loop locations detected in [18] coincided with genes, and there is evidence that R-loops form concurrently with transcription. In a process that is yet to be

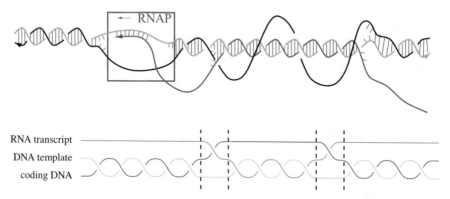

Fig. 1 A schematic depiction of an R-loop. Top: The DNA duplex is formed by two DNA strands in black and green. The black strand represents the coding DNA strand, while the green represents the DNA template (non-coding). The red strand represents the RNA transcript. The 3′-ends are indicated with an arrowhead. The blue box assumes the polymerase reading of the template DNA, and synthesizing the RNA. Bottom: Simplified depiction of the R-loop. We assume that this diagram is read from left to right, with the polymerase on the left (outside the image). The region between the two leftmost vertical dashed lines indicates the location where the RNA transcript invades the DNA duplex, thus initiating the R-loop. Likewise, the two vertical dashed lines on the right indicate the R-loop termination region

understood, the RNA transcript occasionally hybridizes with the DNA template and the second (coding) DNA strand 'entangles' with the RNA:DNA duplex causing the formation of a *co-transcriptional R-loop* [2].

Transcription and the effect of DNA topology on R-loop formation. Transcription is a molecular process that converts a gene encoded in a double stranded DNA molecule into RNA transcript, which is eventually translated into a protein. The double-stranded DNA consists of two sugar-phosphate backbones lined up by complementary sequences of nucleotides (A,T,C,G). One of the strands is the coding strand (i.e. it carries genetic code) and is indicated in black in Fig. 1. The other strand, complementary to the coding strand, is the template strand (indicated in green in Fig. 1). The DNA template is transcribed into RNA by the RNA polymerase. The coding and template DNA strands form a double helix held together by hydrogen bonds. Therefore the transcription machinery (blue box in Figs. 1 and 2a) must break the bonds and open the helix before the RNA polymerase can use the template DNA to produce the complementary RNA transcript. Because DNA is right-handed, the opening of the helix induces an accumulation of torsional stress due to over twisting.

Over twisting is promptly converted into positive supercoiling ahead of the RNA polymerase and compensatory negative supercoiling behind (see twin supercoiling domain, Fig. 2). Note that the local accumulation of supercoiling during transcription, added to the presence of any ambient supercoiling of the DNA, increases torsional stress on the DNA duplex. These factors play a role during the branch migration involving DNA template dissociation from its complementary DNA strand (the coding DNA) and hybridizing with the newly formed RNA transcript, forming and stabilizing an R-loop. To learn more about DNA topology and the effect of transcription in this context, the reader is referred to [1]. The field of DNA topology studies the topology (e.g., knotting and spatial embeddings) and geometry (e.g., twisting, supercoiling) of circular or topologically constrained DNA molecules.

An energy-based statistical mechanical model of R-loop formation was proposed in [21]. This model and its computer implementation (R-looper) incorporates contributions from both sequence and DNA topology to predict the most favorable locations of R-loop formation assuming that the system is in equilibrium. R-looper aims to identify factors that contribute to changes in energy during R-loop formation, and to predict genetic locations favorable for R-loops. The strong role of DNA supercoiling in R-loop formation predicted by the simulation was experimentally supported, as was the strong role of the intrinsic properties of the template DNA sequence (specifically, its *C-richness* and *CG-skew*) [6, 9, 15, 16, 24]. Based on these results a study of R-loop formation must include a discussion of both topology and sequence contributions [21].

In Sect. 2, we introduce a model for R-loops based on a formal grammar with the goal of building a framework for describing the structure of the R-loops and the spatial molecular embedding. This model sets the stage for a future, more sophisticated grammar that could take into account topology and geometry of R-loops (see discussion in Sect. 5). In formal language theory, a grammar is a set of

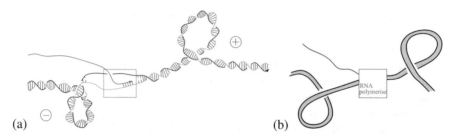

Fig. 2 Local changes in DNA topology during transcription: the twin supercoil domain. The term DNA topology is used by biologists to refer to both topology and geometry of DNA. A supercoil corresponds to a crossing of the axis of the DNA double-helix over itself. When the axis is assigned an orientation, the supercoils are positive or negative depending on the sign of the crossing (as indicated in the image). (**a**) As the DNA template (in green) is transcribed into RNA (in red), positive (+) supercoils accumulate ahead of the polymerase and compensatory negative (−) supercoils accumulate behind. The arrow on the (red) RNA strand indicates the direction of the polymerase. (**b**) The duplex DNA is represented as a ribbon, omitting the helical twists, showing only the positive and the negative supercoils

production rules that generate strings in a formal language. Applications of formal grammars can be found in a wide range of areas from theoretical computer science, to theoretical linguistics, to molecular biology. In molecular biology, applications include modeling regulation of gene expression [3], gene structure prediction [4], and RNA secondary structure prediction [17].

The formal grammar model for R-loops presented here focuses on the structure of an R-loop as described by the braiding of the strands as illustrated in Fig. 1 (bottom), and is informed by sequence contributions. More precisely, the proposed grammar rules depend on the relative nucleotide sequence favorability for R-loop formation. Several experimental results indicate that the presence of a G-rich RNA transcript provides relatively higher thermodynamic stability of an RNA:DNA duplex over a DNA:DNA duplex [10, 13, 14, 22]. This may lead to the breaking of the hydrogen bonds within a topologically strained DNA:DNA duplex. Breakage can then trigger a RNA:DNA branch migration, and the affinity for hybridization of a G-rich RNA transcript with its DNA template may yield favorable regions for R-loop formation. We propose a test for the sequence dependency for R-loop formation and compare our approach with the results from R-looper whose results have been experimentally tested with two plasmids [21].

This chapter is organized as follows. Section 2 gives the necessary background on formal grammars. Section 3 defines the grammar that describes a language for R-loops. Section 4 discusses incorporating the nucleotide sequence dependency into the mathematical framework. In particular, our model takes into account sequence contributions from C-richness and CG-skew. We conclude with an outline of future steps, including a discussion on the contributions of DNA topology and other entanglement considerations in Sect. 5.

2   Formal Grammars: An Overview

In this section we give a short background on formal grammars needed for this work. Good introductions to the different types grammars and languages as well as their properties can be found in [8] and [20].

A finite set Σ is called an *alphabet* and its elements are called *symbols*. For an alphabet Σ, let Σ^* denote the set of all finite sequences of symbols called *strings* or *words* formed by the symbols of Σ. The *empty word* is a word with no symbols and is denoted ϵ. We set $\Sigma^+ = \Sigma^* \setminus \{\epsilon\}$. For example, if $\Sigma = \{a, b\}$, then the set of words over Σ is $\Sigma^* = \{\epsilon, a, b, aa, ab, ba, bb, aaa, aab, aba, \dots\}$ while $\Sigma^+ = \{a, b, aa, ab, ba, bb, aaa, aab, aba, \dots\}$. We use lower case letters at the beginning of the Roman alphabet (e.g., $a, b, c, \dots$) to indicate symbols and lower case letters at the end of the Roman alphabet to indicate words (e.g., $u, v, w, \dots$).

For a word $u = a_1 a_2 \cdots a_n$, where $a_i \in \Sigma$, we say that the *length* of u is $|u| = n$. The length of the empty string ϵ is 0. The number of symbols in u equal to a is denoted $|u|_a$, for example $|aab|_a = 2$ and $|aab|_b = 1$. For any $1 \le i < j \le n$, the substring $a_i a_{i+1} \cdots a_j$ of u is denoted by $u_{[i,j]}$.

Definition 1 A *grammar* Γ is a 4-tuple (S, N, Σ, P), where

- N is an alphabet whose symbols are called *nonterminals*,
- Σ is an alphabet whose symbols are called *terminals*,
- P is a finite set of *production rules* of the form $w \to w'$ for some words $w, w' \in (\Sigma \cup N)^*$ provided that at least one symbol in w is nonterminal, and
- $S \in N$ is a designated nonterminal called the *start symbol*.

Here we adopt the standard convention where upper case Roman characters (e.g. A, B, S) are used to denote the nonterminals, and lower case Roman characters (e.g. a, b, c) to denote the terminals. Let u be a word in $(\Sigma \cup N)^*$ and $r : w \to w'$ be a rule in P. Applying the production rule r to the word (or string) u means finding a substring w in u and replacing it with w', while keeping the rest unchanged. For example, if $x, y, w \in \Sigma^*$, applying rule $w \to w'$ to the word $u = xwy$ produces $v = xw'y$. We write $u \overset{r}{\Rightarrow} v$.

We say that a word $w \in \Sigma^*$ can be *derived*, and denote it as $S \overset{*}{\Rightarrow} w$, if there is a sequence of production rules $r_1, r_2, \dots, r_n$ and a sequence of words in $w_1, \dots, w_n \in (\Sigma \cup N)^*$, where $w_n = w$, such that

$$S \overset{r_1}{\Rightarrow} w_1 \overset{r_2}{\Rightarrow} w_2 \overset{r_3}{\Rightarrow} \cdots \overset{r_n}{\Rightarrow} w_n = w.$$

Such a sequence of applications of rules $r_1, r_2, \dots, r_n$ is called a *derivation* of w. The *language* described by the grammar Γ is the set of all words with only terminal symbols that can be derived, i.e.,

$$L(\Gamma) := \{w \in \Sigma^* \mid S \overset{*}{\Rightarrow} w\}.$$

Example 1 Consider the grammar $\Gamma = (S, N, \Sigma, P)$ with

$$N = \{S, A, B\} \qquad \Sigma = \{a, b\}$$
$$P = \{r_1 : S \rightarrow aA, \; r_2 : S \rightarrow bB, \; r_3 : A \rightarrow aA,$$
$$r_4 : B \rightarrow bB, \; r_5 : A \rightarrow \epsilon, \; r_6 : B \rightarrow \epsilon, \; r_7 : S \rightarrow \epsilon\}.$$

The word *aaaa* is in $L(G)$ because it can be derived in the following way:

$$S \overset{r_1}{\Rightarrow} aA \overset{r_3}{\Rightarrow} aaA \overset{r_3}{\Rightarrow} aaaA \overset{r_3}{\Rightarrow} aaaaA \overset{r_5}{\Rightarrow} aaaa.$$

Based on its production rules, the language $L(\Gamma)$ described by this grammar Γ consists of words with a single symbol, that is

$$L(\Gamma) = a^* \cup b^*.$$

We note that a common abuse of notation when the alphabet is just a singleton is to replace $\{a\}$ with a, and $\{a\}^*$ with a^*.

We often remove the superscript over the arrows when rules are easily identifiable and write $u \rightarrow v$ instead of $u \overset{r}{\Rightarrow} v$. We use the convention to shorten the description of the rules that have the same left hand-side by using vertical bars. For example, the expression $w \rightarrow w_1 \mid w_2 \mid w_3$ means that the set of rules P contains $w \rightarrow w_1$, $w \rightarrow w_2$, and $w \rightarrow w_3$. With these conventions, a grammar is completely determined by the list of production rules.

Example 2 Consider the grammar defined with rules:

$$S \rightarrow aSb \mid \epsilon.$$

In this grammar there are two rules, $S \rightarrow aSb$ and $S \rightarrow \epsilon$. The symbol S is the only nonterminal, and the set of terminals is $\{a, b\}$. Then, every derivation is of the form:

$$S \Rightarrow aSb \Rightarrow aaSbb \cdots \Rightarrow aa \cdots aSbb \cdots b \Rightarrow aa \cdots abb \cdots b.$$

Because with the application of the rules the number of a's remains equal to the number of b's, the language defined by this grammar is $L = \{a^n b^n \mid n \geq 0\}$.

In formal language theory, the Chomsky hierarchy refers to the containment hierarchy of four levels of languages: regular, context-free, context-sensitive and computably enumerable languages [20].

A grammar is said to be *regular* if all production rules are of the type: $A \rightarrow a$, $A \rightarrow aB$, or $A \rightarrow \epsilon$, where $A, B \in N$ and $a \in \Sigma$. This means that with each rule, either a nonterminal is replaced by a terminal, or it is replaced by a terminal followed by another nonterminal, or it is erased. *Regular languages* are those that can be generated by a regular grammar. The grammar in Example 1 is regular and, consequently it describes a regular language $a^* \cup b^*$.

In *context-free* grammars, the rules are of the type $A \rightarrow x$, where $A \in N$ and x is a string (possibly empty) of terminals and nonterminals. In *context-sensitive* grammars, the rules are of the type $xAy \rightarrow xzy$, where x, y, z are strings of terminals and nonterminals. Context-free (respectively, context-sensitive) languages are the ones that can be generated by context-free (respectively, context-sensitive) grammars. The grammar in Example 2 and the language it generates are context-free. One can show that this language is not regular, i.e., there is no regular grammar that defines this language [8, 20]. *Computably enumerable languages* are defined by grammars without constraints.

3 R-Loop Grammars

An R-loop is a structure consisting of a RNA:DNA hybrid and a displaced DNA single strand (Fig. 1). While here we focus on co-transcriptional R-loops, the mathematical framework can be applied to any R-loop or other nucleic acid triplex. First we summarize the transcription process which infers the construction of the grammar.

During transcription, the RNA polymerase complex binds to the promoter region of a gene in a double-stranded DNA molecule, unwinds the DNA double helix, and transcribes the template strand into a single-stranded RNA molecule (the *RNA transcript*) one nucleotide at a time in the $3' \rightarrow 5'$ direction. The nucleotide sequence of the RNA transcript is complementary to that of the DNA template, and is identical to the sequence of the coding DNA after replacing each T with a U. As transcription proceeds along the template (as the polymerase moves) the RNA transcript exits the 'bubble' formed by the polymerase complex and the unwound DNA duplex. Simultaneously, the DNA double helix reforms behind the complex. At the end of the process the RNA transcript is released. For reasons that are still unclear, occasionally the RNA transcript hybridizes with the DNA template thus giving rise to a co-transcriptional R-loop.

We represent the formation of an R-loop as a string (word) over the alphabet $\Sigma = \{\sigma, \hat{\sigma}, \tau, \hat{\tau}, \alpha, \omega\}$. Each symbol in the alphabet can be described as a 3-stranded local structure corresponding to the length of one half turn of B-form DNA, approximately 5 nucleotides (see Fig. 3). The symbols τ and $\hat{\tau}$ represent a RNA:DNA hybrid, σ and $\hat{\sigma}$ represent a DNA:DNA duplex, and α and ω represent a structure where all three strands interact. Note that in τ and $\hat{\tau}$, and in σ and $\hat{\sigma}$, the third strand is assumed to not interact (via hydrogen bonds) with the duplex.

Presence of $\hat{\ }$ on top of the symbols, such as $\hat{\sigma}$ or $\hat{\tau}$, indicates that the corresponding duplex is in a stable configuration. Less stable half-turn configurations are denoted by σ and τ, without the $\hat{\ }$, are more likely to transition into one of the 3-stranded hybrids α or ω via strand branch migration. The production rules will be guided by the stability of the half-turns σ and τ (DNA:DNA and RNA:DNA,

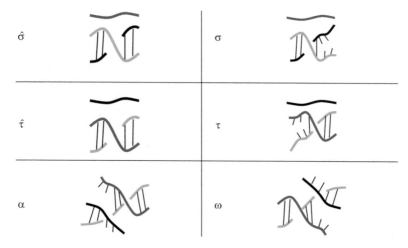

Fig. 3 Interpretation of each of the symbols used in the grammar. The black strand represents the coding DNA strand, the green strand represents the template DNA strand, and the red strand represents the RNA transcript. Here, σ and $\hat{\sigma}$ are DNA:DNA hybrids, τ and $\hat{\tau}$ are RNA:DNA hybrids, and α and ω are transitions between the two. The ' ^ ' indicate more stable configurations. Less stable configurations are depicted with some breakage in the hydrogen bonds to suggest that there is more prevalent 'breathing' of the duplex in that region. The breakage in σ and τ is indicated only by symbols in Fig. 4

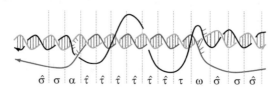

Fig. 4 An example R-loop associated with the word $\cdots \hat{\sigma}\sigma\alpha\hat{\tau}\hat{\tau}\hat{\tau}\hat{\tau}\hat{\tau}\hat{\tau}\tau\omega\hat{\sigma}\sigma \cdots$. Note that if the sequence stability weakens within an R-loop then a τ may follow after an initial string of one or more $\hat{\tau}$'s, and this may lead into an R-loop termination region, indicated by ω. The three strand sections corresponding to α and ω indicate the branch migration when RNA 'invades' the DNA duplex (α) and leaves the duplex (ω). Observe that there may be other words that correspond to the same R-loop, because the sequence stability may vary both within and outside the R-loop

respectively). The remaining two symbols of the alphabet, α and ω, are used to represent R-loop initiation and termination regions. The start of the R-loop, denoted by α, is the structure formation at the moment when RNA 'invades' the DNA:DNA duplex. The end of the R-loop, denoted by ω, is the structure obtained when the RNA dissociates from the RNA:DNA duplex and the DNA returns to its native state. Figure 4 illustrates how an R-loop can be represented by a string over Σ.

How the different symbols are assigned to the specific genomic region will be guided by the biology. Some nucleotide sequences that are prone to R-loop formation have been identified experimentally and models have been proposed to predict them. As a first approximation, in the next section we use the preliminary data used in a recent energetics model to assign stable vs unstable half-turn segments in both a RNA:DNA duplex and in a DNA:DNA duplex [21]. We assume that a DNA sequence of nucleotides is favorable for R-loop formation (i.e. the RNA:DNA duplex is more stable) if it is C-rich and CG-skewed (see Sect. 4 for definitions of these terms).

We assume that an R-loop necessarily starts with a short nucleotide sequence that has an "unstable" DNA:DNA duplex, indicated by σ. The three strand formation when the RNA 'invades' the DNA duplex through branch migration is indicated by α and is followed by a half-turn of a stable RNA:DNA duplex indicated by $\hat{\tau}$. Hence the word contains the subword $\sigma\alpha\hat{\tau}$. Similarly, the end of the R-loop is obtained from a sequence starting with at least one unstable RNA:DNA half-turn denoted by τ, followed by the three strand formation ω where the RNA dissociates from the DNA and by a stable DNA duplex ($\hat{\sigma}$). Hence the R-loop word must also contain the subword $\tau\omega\hat{\sigma}$.

The following grammar Γ generates the words associated with R-loops. Recall from Sect. 2 that by defining the production rules we uniquely define a grammar.

the grammar Γ :

$$S \rightarrow \hat{\sigma}D \mid \sigma D \qquad \text{rules:start}$$
$$\hat{\sigma}D \rightarrow \hat{\sigma}\hat{\sigma}D \mid \hat{\sigma}\sigma D \qquad \text{rules:s} - \text{D} - \text{duplex}$$
$$\sigma D \rightarrow \sigma\hat{\sigma}D \mid \sigma\sigma D \mid \sigma\alpha\hat{\tau}R \qquad \text{rules:us} - \text{D} - \text{duplex}$$
$$\hat{\tau}R \rightarrow \hat{\tau}\hat{\tau}R \mid \hat{\tau}\tau R \qquad \text{rules:s} - \text{R} - \text{duplex}$$
$$\tau R \rightarrow \tau\hat{\tau}R \mid \tau\tau R \mid \tau\omega\hat{\sigma}D' \qquad \text{rules:us} - \text{R} - \text{duplex}$$
$$D' \rightarrow \hat{\sigma}D' \mid \sigma D' \mid \epsilon \qquad \text{rules:end}$$

The set of nonterminals in Γ is $N = \{S, D, R, D'\}$, where each of the nonterminals is associated with one of the hybrid structures. The nonterminals D and D' are used to generate DNA duplexes before and after the R-loop, respectively. The nonterminal R is used to generate the symbols that correspond to the RNA:DNA duplex within the R-loop. The symbol S is the starting symbol of the grammar. The grammar Γ uses six types of rules as indicated above. Rules start are the starting rules that generate either σD or $\hat{\sigma}D$, a half-turn DNA duplex. If the DNA duplex represented by $\hat{\sigma}$ is stable, then according to rules s-D-duplex (stable DNA duplex) the next half-turn must be another DNA duplex, which could be stable ($\hat{\sigma}D$), or not (σD). If the DNA duplex represented by σ is not stable (i.e. σD), then according to the rules us-D-duplex (the unstable duplex rules) it can also be followed by a three strand formation α and a stable RNA:DNA duplex $\hat{\tau}$ (e.g. $\sigma D \rightarrow \sigma\alpha\hat{\tau}R$). Rules s-R-duplex (stable RNA:DNA duplex) and rules us-R-duplex (unstable RNA:DNA duplex) are analogous to the rules s-D-duplex and us-D-duplex, except that they generate the string corresponding to the R-loop. The first two rules in end are analogous to rules start, with the addition of the last rule $D' \rightarrow \epsilon$ which is used to stop the word derivation.

The grammar Γ is context-sensitive, meaning that the rules with nonterminals D and R on the left hand side depend on the preceding symbol. Recall that the language derived from the grammar Γ is defined as the set of words with only terminal symbols which can be derived. Describing the set of words generated by the grammar is straightforward. Every word derivation in Γ starts with S and generates σ or $\hat{\sigma}$ followed by a nonterminal D. Depending on whether σ or $\hat{\sigma}$ precedes D, the next symbols that are generated are again σ or $\hat{\sigma}$ (rules s-D-duplex). In addition, if σ precedes D (rules us-D-duplex), then the next symbols could be $\sigma\alpha\hat{\tau}$ which generate the word $x\sigma\alpha\hat{\tau}R$ where $x \in \{\sigma, \hat{\sigma}\}^*$. After this word one can only apply rules s-R-duplex which generate new symbols τ or $\hat{\tau}$ with a non-terminal R. Rules s-R-duplex and us-R-duplex are then applied to symbols $\hat{\tau}$ and/or τ followed by R. If at some point we use the last rule of us-R-duplex, the symbols that follow are $\tau\omega\hat{\sigma}$ followed by a nonterminal D', and the corresponding word has the form $x\sigma\alpha\hat{\tau}y\tau\omega\hat{\sigma}D'$, where $y \in \{\tau, \hat{\tau}\}^*$. Note that once D' appears in a word, the rules end are the only rules that can be applied and they generate symbols σ or $\hat{\sigma}$. The derivation stops with an application of rule $D' \to \epsilon$. In sum, the final word generated by the grammar has the form $x\sigma\alpha\hat{\tau}y\tau\omega\hat{\sigma}z$ where $z \in \{\sigma, \hat{\sigma}\}^*$. Based on this we have the following proposition. We call the formal language specified with Γ the *R-loop language*.

Proposition 1 *The R-loop language described by Γ is*

$$L(\Gamma) = \left\{x\sigma\alpha\hat{\tau}y\tau\omega\hat{\sigma}z \mid x, z \in \{\sigma, \hat{\sigma}\}^*, y \in \{\tau, \hat{\tau}\}^*\right\}$$

or equivalently, $L(\Gamma) = (\sigma \cup \hat{\sigma})^* \sigma \alpha \hat{\tau} (\tau \cup \hat{\tau})^* \tau \omega \hat{\sigma} (\sigma \cup \hat{\sigma})^*$.

As a consequence of Proposition 1, the language of the grammar Γ is regular, that is, it can be described by a regular expression [20]. Therefore, there is a regular grammar $\hat{\Gamma}$ with the same alphabet Σ that is equivalent to Γ. $\hat{\Gamma}$ defines the same R-loop language with the following production rules:

$$S \to \sigma S \mid \hat{\sigma} S \mid \sigma Q_1$$
$$Q_1 \to \alpha Q_2$$
$$Q_2 \to \hat{\tau} Q_3$$
the grammar $\hat{\Gamma}$: $$Q_3 \to \tau Q_3 \mid \hat{\tau} Q_3 \mid \tau Q_4$$
$$Q_4 \to \omega Q_5$$
$$Q_5 \to \hat{\sigma} Q_6$$
$$Q_6 \to \sigma Q_6 \mid \hat{\sigma} Q_6 \mid \epsilon$$

Although $\hat{\Gamma}$ is regular, i.e., it is a different grammar from the initial context-sensitive grammar Γ, the meaning of the symbols σ, $\hat{\sigma}$, τ, $\hat{\tau}$ that indicate nucleotide stability remains unchanged.

4 A Discrete Model to Estimate R-Loop Favorability

In order for the grammar Γ presented in Sect. 3 to be useful, it requires information about how favorable or unfavorable a particular stretch of DNA is for R-loop formation. The grammar as constructed in the previous section contains no such information. We are interested in incorporating into the grammar Γ information on nucleotide sequence contributions combined with the topological changes, to detect and predict R-loop formation. We initiate this line of work in this publication. In this section we start with a simple discrete model for estimating R-loop favorability based only on properties of the nucleotide sequence. More specifically, we focus on two different measures, C-richness (cytosine-rich sequence) and CG-skew (cytosine-guanine ratio).

Intuitively, one would expect R-loops to occur infrequently. However, they account for up to 5% of the human genome, and represent the most common non-B DNA structures quantified to date (reviewed in [2]). General investigations of nucleic acid hybrid stability have measured the relative free energy of a nucleic acid duplex (denoted by $\Delta\Delta G^0$ in kcal/mole) as the free energy of the given duplex minus the free energy of the most stable duplex [13, 14]. The results of these studies are summarized in Fig. 5 and show that a hybrid consisting of a purine-rich RNA and pyrimidine-rich DNA (denoted r(GA).d(CT)) is significantly more stable than the corresponding DNA:DNA hybrid (denoted d(GA).d(CT)). The stabilities of the other two possible duplexes (denoted r(GU).d(CA) and d(GT).d(CA)) are comparable to each other. This suggests that the strand migration initiated by an RNA strand as it invades a DNA:DNA duplex is more likely to occur when the template DNA is C-rich (or equivalently, when the transcript RNA is G-rich). On the other hand, the d(GA).r(CU) stacking is significantly less stable than the d(GA).d(CT) stacking, while the stabilities of r(CA).d(GT) and d(CA).d(GT) (not shown in Fig. 5) are comparable to each other. So, a G-rich DNA template would result in a thermodynamically unfavorable RNA:DNA duplex. Based on this, we expect that the R-loops are more likely to form in regions where the template DNA is C-rich and CG-skew. In the discrete model below we focus on those two quantities.

Experimental observations suggest that C-rich regions that are also CG-skewed are correlated with R-loop occurrence [6]. However, not all CG-skewed areas are associated with R-loops [15]. In [21] the authors proposed a statistical mechanics model of R-loop energetics that provides predictions of R-loop favorability for a given nucleotide sequence. This model takes into account contributions of both the sequence and the supercoiling. The theoretical predictions were tested experimentally on two plasmids [2, 21].

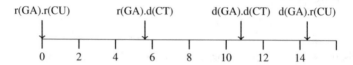

Fig. 5 Plot of the relative stability of duplexes; from the most stable duplex (r(GA).r(CU)) at the left, to the least stable duplex (d(GA).r(CU)) at the right. Stability is determined by computing the relative free energy ($\Delta\Delta G^0$) of the duplexes in kilocalories per mole [13, 14]. The scale indicates the relative free energy of a duplex with respect to the most stable duplex shown as reference at 0. Other duplexes, such as r(CA).d(GT) and d(CA).d(GT) that are above 8 are not indicated in order to keep the figure readable, and because they are not significant in our sequence analysis of R-loops

Before we proceed we introduce several definitions. The DNA alphabet is $\Sigma_{DNA} = \{A, G, C, T\}$, and the binary alphabet is $\Sigma_B = \{0, 1\}$. A DNA sequence is a word $w \in \Sigma_{DNA}^*$. Recall that the number of symbols in w equal to a symbol a is denoted $|w|_a$.

Definition 2 ([7]) The *CG-skew* of a DNA sequence w is a function $C_{sk} : \Sigma_{DNA}^+ \rightarrow [-1, 1]$ where

$$C_{sk}(w) = \begin{cases} \dfrac{|w|_C - |w|_G}{|w|_C + |w|_G} & \text{for } |w|_C + |w|_G > 0 \\ \\ 0 & \text{otherwise .} \end{cases}$$

The CG-skew of a DNA sequence measures the dominance in occurrences of cytosine with respect to guanine. If $C_{sk}(w) = 1$, then the DNA segment w is a sequence that contains cytosines but no guanines. Conversely, $C_{sk}(w) = -1$ indicates a sequence that contains guanines but no cytosines. Further, a 2:1 ratio of cytosine to guanine in a strand would result in a CG-skew value of 0.33.

Definition 3 The *C-richness* of a DNA sequence w is a function $C_r : \Sigma_{DNA}^+ \rightarrow [0, 1]$ where

$$C_r(w) = \frac{|w|_C}{|w|}.$$

In order to explore the contributions of C-richness and CG-skew as predictors for R-loops, we consider the plasmids (given as template strands in the 5'-3' direction) of the experimental analysis from the Chédin lab [21].

We compute CG-skew and C-richness using a sliding window approach. Given thresholds t_1 and t_2, and a sequence of nucleotides, we associate a binary score for

CG-skew and C-richness to each subword of length ℓ; a score of 1 if the threshold is met, and a score of 0 if the threshold is not met. We employ the following definitions.

Let $f : \Sigma_{DNA}^+ \to \mathbb{R}$ be a function defines on the set of DNA sequences (i.e. Σ_{DNA}^+). The function can be measuring properties of the sequence, such as C-richness or CG-skew. For each function f let t be a real number indicating the threshold. A ℓ-*window t-threshold for* f is the function $T_{\ell,f}^t : \Sigma_{DNA}^+ \to \Sigma_B$ defined with $T_{\ell,f}^t(w) = b_1 b_2 \cdots b_{|w|}$ where

$$\text{for } 1 \leq i \leq |w| - \ell \qquad b_i = \begin{cases} 1 \text{ if } f(w_{[i,i+\ell-1]}) \geq t \\ 0 \text{ otherwise} \end{cases}$$

$$\text{for } |w| - \ell < i \leq |w| \qquad b_i = \begin{cases} 1 \text{ if } f(w_{[i,|w|]}) \geq t \\ 0 \text{ otherwise.} \end{cases}$$

Note that as we reach the end of the string, if there are fewer than ℓ nucleotides left, we consider the same thresholds but on the remaining, shorter, substrings. Thus it becomes more difficult to meet the predetermined thresholds towards the end of the string.

Of interest to the study of R-loops are two functions: the ℓ-window t_1-threshold for CG-skew, $T_{\ell,C_{sk}}^{t_1}$; and the ℓ-window t_2-threshold for C-richness, $T_{\ell,C_r}^{t_2}$. Starting from the first nucleotide in the template DNA we consider a substring of length ℓ and compute the values of $T_{\ell,C_{sk}}^{t_1}$ and $T_{\ell,C_r}^{t_2}$ moving from left to right one nucleotide at each step. If a substring is found to be both CG-skewed and C-rich, then we say it is *double-C-rich*. The double-C-rich string corresponding to w is $w_B = T_{\ell,C_{sk}}^{t_1}(w) \wedge T_{\ell,C_r}^{t_2}(w)$. For two binary strings $u = b_1 \cdots b_k$ and $u' = b_1' \cdots b_k'$ we define $v = u \wedge u' = c_1 \cdots c_k$ such that for all $i = 1, \ldots, k$ we have $c_i = 1$ if and only if $b_i = b_i' = 1$.

We create a binary string of double-C-richness for the entire nucleotide string (called the *double-C-rich string*). In Examples 3–4, we illustrate these definitions for a hypothetical string of nucleotides.

Example 3 Consider an example string of length 30 in Σ_{DNA}^+ given by $w =$ AGAGCCCGATCCAGACCCCGACGTTACGAA and a window size $\ell = 10$. Suppose the CG-skew threshold is $t_1 = 0.3$ and the threshold for C-richness is $t_2 = 0.5$.

In the first ten nucleotides, there are 3 C's and 3 G's, so the CG-skew $C_{sk}(w_{[1,10]})$ is $\frac{0}{6}$. For nucleotides 2–11 the CG-skew is $C_{sk}(w_{[2,11]}) = \frac{1}{7}$ and for nucleotides 3–12 we have $C_{sk}(w_{[3,12]}) = \frac{3}{7}$. Since $\frac{3}{7} \geq 0.3$, the first three symbols in the ℓ-window t_1-threshold for $T_{\ell,C_{sk}}^{t_1}(w)$ are 001. The entire string is

$$T_{10,C_{sk}}^{0.3}(w) = 001111111111101111000000000000.$$

If the threshold changes to $t_1 = 0.35$, the string becomes

$$T^{0.35}_{10,C_{sk}}(w) = 001110001111101100000000000000.$$

In this example the threshold for C-richness is $t_2 = 0.5$, and there are only 3 C's in the first ten nucleotides, then the C-richness is $C_r(w_{[1,10]}) = \frac{3}{10} < 0.5$. Therefore the first ten nucleotides are not C-rich and the first symbol of $T^{t_2}_{\ell,C_r}(w)$ is 0. Nucleotides 3–12 have 5 C's, so this meets the threshold of 0.5 and is considered a C-rich region. The resulting string with threshold 0.5 is $T^{0.5}_{10,C_r}(w) = 001110001111111100000000000000$. If we use a threshold of 0.6, the string becomes $T^{0.6}_{10,C_r}(w) = 000000000110000000000000000000$.

Only when the window of size ℓ is both CG-skewed and C-rich do we say that it is double-C-rich, and the corresponding entry in the string w_B receives a value of 1. With thresholds 0.3 and 0.5 (for CG-skew and C-richness, respectively), $w_B = T^{0.3}_{10,C_{sk}}(w) \wedge T^{0.5}_{10,C_r}(w) = 001110001111101100000000000000$. Observe that the string w_D is sensitive to the thresholds chosen. With thresholds $t_1 = 0.3$ and $t_2 = 0.6$, for the same nucleotide sequence the string becomes $w_B = T^{0.3}_{10,C_{sk}}(w) \wedge T^{0.6}_{10,C_r}(w) = 000000000110000000000000000000$.

We consider that an isolated occurrence of double-C-richness does not correspond to likelihood of R-loop formation. This agrees with an in vitro analysis that showed that some accumulation of double-C-richness provides an optimal situation for R-loop formation [15]. Therefore here we consider an *accumulation string with window j* to be a function $Acc^j : \Sigma_B^+ \to \{1, \ldots, j\}^*$ defined by $Acc^j(w_B) = w_A = a_1 a_2 \cdots a_k$ where

$$a_i = \begin{cases} |w_{B[i,i+j-1]}|_1 & \text{for } 1 \leq i \leq |w_B| - j \\[2mm] |w_{B[i,|w_B|]}|_1 & \text{for } |w_B| - j < i \leq |w_B|. \end{cases}$$

The string $w_A = Acc^j(w_B)$ is obtained from the binary string $w_B = b_1 b_2 \cdots b_k$ such that a_i gives the number of 1's within a window of size j in the substring $b_i b_{i+1} \cdots b_{i+j-1}$. For any fixed pair of thresholds t_1 and t_2 of C-richness and CG-skew, the values of the symbols in the accumulation string can be interpreted as an indication of R-loop likelihood.

Example 4 Consider the sequence w from Example 3. If we record occurrences of double-C-richness within a window size $j = 5$, then the double-C-rich binary string $w_B = 001110001111101100000000000000$ has an accumulation string of $Acc^5(w_B) = w_A = 333222345444322100000000000000$. Note that $a_1 = 3$, since the substring $b_1 \cdots b_5 = 00111$ has three 1's.

Remark 1 The accumulation strings give a sequence-based estimate for favorable sites for R-loop formation. Such portions of the accumulation strings with high values help in deciding which rule of Γ applies for each half turn of DNA. The accumulation string in Example 4 has length 30 and corresponds to six half-turns of DNA (five nucleotides per half-turn) and therefore it can be represented with a word containing six σ,τ-symbols. If, for example, we set a threshold of 4 as a minimum requirement in the accumulation string for R-loop favorability, then the length 5 substrings in the accumulation string that contain a 4 (or larger) could correspond to symbols σ, $\alpha\hat{\tau}$, or $\hat{\tau}$, while the length 5 substrings with no values larger than 3 could correspond to symbols τ, $\omega\hat{\sigma}$, or $\hat{\sigma}$. We note that symbols α and ω indicate the location of the 3-strand branch migration. For simplicity we assume that they correspond to transitions between a DNA:DNA hybrid and a RNA:DNA hybrid, and do not assign a number of nucleotides to them.

In Example 4 the segment $a_1 a_2 \ldots a_5 = 33322$ of the string w_A does not correspond with a favorable region. In this case the first symbol of a word in the R-loop language corresponding to w would be $\hat{\sigma}$. Since $a_8 = 4$, $a_9 = 5$ and $a_{10} = 4$, the second symbol of the word (corresponding to $a_6 \cdots a_{10} = 23454$) could be σ. The string $a_{11} \cdots a_{15} = 44322$ continues to be favorable for R-loop formation, and could in fact indicate the start of an R-loop. Continuing in a similar manner, one possibility for a word from the grammar corresponding to our accumulation string would be $w_R = \hat{\sigma}\sigma\alpha\hat{\tau}\tau\omega\hat{\sigma}\hat{\sigma}$. The use of a threshold of 4 for R-loop favorability in this example is only for illustration purposes.

Hence, starting from a DNA string w, we apply ℓ-window threshold functions for C_{sk} and C_r to obtain the corresponding double-C-rich binary string w_B, to which we associate a j-window accumulation string w_A. Using an accumulation threshold we then produce a word w_R in the R-loop language providing structural information about the DNA:DNA and RNA:DNA hybrids and branch migration.

In order to compare the likelihood of R-loop formation for a given DNA string w using its accumulation string w_A to the existing energetics model [21], we obtained the template sequences (written in the 5' to 3' direction) for the experimentally tested plasmids $w = $ pFC53 of length 3906 nucleotides, and $w' = $ pFC8 of length 3669 nucleotides (personal communication with the authors). The window value for $T^{t_1}_{\ell, C_{sk}}$ and $T^{t_2}_{\ell, C_r}$ for both plasmids was set to $\ell = 10$. We examined the sequences for both plasmids, and experimented with different ℓ-window threshold values t_1 and t_2 of CG-skew and C-richness. For each pair (t_1, t_2), we created an accumulation string w_A by recording the number of occurrences of double-C-richness in each window of length j. We tested values of j between 5 and 100 and found that when j is too small, the accumulation strings are not sensitive enough, and when j is too large, the values in w_A and w'_A miss fluctuations within the double-C-rich regions. We used $j = 50$ (this roughly corresponds to 5 full turns of DNA) for the strings w and w' associated with the plasmids pFC53 and pFC8. We considered all possible pairs of threshold values t_1, t_2 between 0 and 1 with step size 0.05 for the functions C_{sk} and C_r (a total of 400 accumulation strings, 20 for each threshold), and computed the accumulation strings w_A and w'_A. We then compared accumulation string w_A and w'_A to the R-looper output probability string (both indexed by nucleotide position number), and computed Spearman's rank correlation coefficient to find the accumulation string with optimal threshold values.

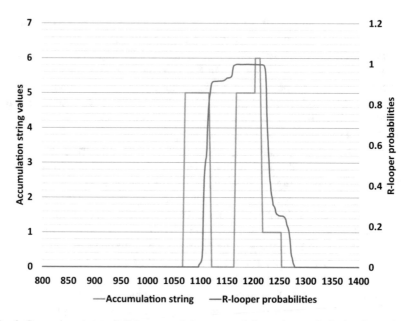

Fig. 6 Comparison between R-looper and the accumulation string analysis for the region of plasmid pFC53 from nucleotide position 800 to 1400. The R-looper probabilities output is indicated by the blue curve. The accumulation string values for pFC53 with $t_1 = 0.1$ and $t_2 = 0.8$ are plotted in orange. Each entry of the accumulation string counts the number of 1's in the corresponding double-C-rich string in the succeeding 50 nucleotides. The region of the plasmid not included in the figure has an R-loop probability zero, and the corresponding accumulation string consists entirely of zeros

For plasmid pFC53 (3906 nucleotides), the correlation coefficient was maximized (at 0.799) when $t_1 = 0.1$ and $t_2 = 0.8$. This result is somewhat surprising since it suggests that C-richness is the determining factor in approximating the R-looper output. The result is shown in Fig. 6. In the graph the maximum accumulation string value was normalized to the same height as a probability of 1 from R-looper.

For plasmid pFC8 (3669 nucleotides) we found optimal thresholds of $t_1 = 0.1$ and $t_2 = 0.6$, with Spearman coefficient of 0.658. Again we see that C-richness is driving the correlation with the R-looper output, as depicted in Fig. 7. The code used for these computations is available publicly at https://github.com/mandariehl/rloopsplusstats.

Comparing the results for these two plasmids, one can observe that the optimal C_r thresholds t_2 disagree. This suggests that C-richness is a larger driving force in R-loop formation than CG-skew, at least when taking R-looper results as a reference. Because the CG-skew threshold is not 0, we believe it is important to continue to include this parameter as structure and topology is incorporated with this model.

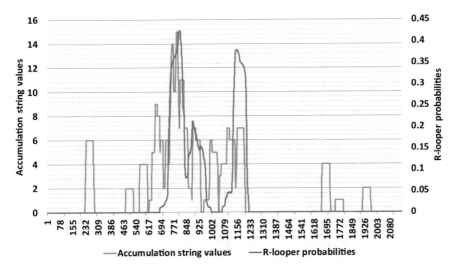

Fig. 7 Comparison between R-looper and the accumulation string analysis for the region of plasmid pFC8 from nucleotide position 1 to 2170. The R-looper probabilities output is indicated by the blue curve. The accumulation string values with $t_1 = 0.1$ and $t_2 = 0.6$ are plotted in orange. Each entry of the accumulation string counts the number of 1's in the corresponding double-C-rich string in the succeeding 50 nucleotides. The plasmid region not included in the figure has R-loop probability zero, and its accumulation string consists entirely of zeros

5 Discussion

The study of R-loops has gained visibility in recent years due to their prevalence and importance for the well-being of the cell. Understanding their mechanism of formation as well as their geometry and topology is key to establishing their biological role. In this paper we propose an abstract framework to model R-loops based on formal grammars, as well as a simple method to assess probability of R-loop formation based on sequence contributions.

It is of interest to obtain a stochastic grammar and a probabilistic model for R-loop formation by attaching probabilities to the production rules in the R-loop grammar Γ proposed in Sect. 3.

Proposition 1 shows that the R-loop language is regular. The class of regular languages is the class of languages described by regular grammars, and is equivalent to the class of languages accepted by finite state automata [20]. Finite state automata equipped with probability measure associated with their state transitions are Markov chain discrete dynamical systems. The framework proposed in Sect. 4 sets the base for determining appropriate probabilities associated with each transition rule in $\hat{\Gamma}$, which is the subject of a future study. Such a model could supplement the predictions based on energy minimization given by R-looper.

A potential advantage of a probabilistic modeling approach is that it can be readily extended to include future sources of statistical information of R-loop formation. An example of a successful probabilistic model is Pfold [11, 12], which combines a stochastic context-free grammar with evolutionary tree information for a consensus secondary structure prediction of homologous RNA sequences.

The design of a probabilistic grammar affects the prediction of R-loop formation. One advantage of 'lightweight' grammars such as the one proposed in this work is the practicality in their implementation. Moreover, the simplicity of the grammar does not necessarily imply poor predictive power, as can be seen in the case of RNA secondary structure prediction [5].

Since our grammar $\hat{\Gamma}$ is regular and each R-loop has exactly one derivation, the probabilities can be obtained by simple counting: one needs to determine the frequency with which each production rule is used for a set of R-loops. On the other hand, since in Γ the symbols σ and τ correspond to a half-turn pairwise interaction between the two DNA strands, for the template DNA and the RNA strand, the derived strings from such a grammar can be used to infer the three dimensional structure of the molecule when the R-loop initiates.

In Sect. 4, we used plasmids pFC53 and pFC8 [21] to test a simple measure of R-loop favorability based on C-richness and CG-skew. In this way, a given sequence is assigned optimal thresholds t_1, t_2 for R-loop formation. The optimal thresholds obtained for the two plasmids turned out to be different. Averaging them results in a drop of the Spearman correlation coefficient to 0.634 for pFC53 and 0.345 for pFC8. This suggests that C_r and C_{sk} values are not sufficient to predict R-loops. Indeed, it has been shown that R-loops are very sensitive to changes in the topology of the DNA template and that negative superhelicity is required for R-loop stability [21]. The parameters of our discrete model are sensitive to the nucleotide sequence and provide useful information about R-loop favorability in some regions and to inform on the eventual probability assignments to grammar rules. In a refinement of the model, one should consider supercoiling and other measures of topological entanglement.

A discussion on entanglement and geometry must include both the topology of the double stranded DNA before and after the formation of the R-loop, as well as a detailed description of the wrapping of the single stranded DNA around the RNA:DNA hybrid. Important considerations that have not been incorporated in the current model include a description of the wrapping of the nontemplate DNA around the RNA:DNA hybrid, and the supercoiling of DNA prior to R-loop formation. It has been observed that superhelicity can have a dramatic effect on R-loop formation [21]. The twin supercoiling domain model (see Fig. 2) predicts that transcription induces positive supercoiling ahead of the transcription complex, and negative supercoiling behind it. When added to the ambient supercoiling of the DNA template, the negative supercoiling behind the polymerase increases the energetic favorability for R-loop formation [21]. It is of interest to use experimental data to expand Γ with probability parameters, as well as to include topological and geometric constraints within the production rules.

Acknowledgments This research was initiated at the Collaborative Workshop for Women in Mathematical Biology at IPAM in 2019. The authors express their gratitude to the AWM and

SIAM for funding, and to IPAM for fostering an exceptional work environment. The authors acknowledge partial support from NSF grants DMS-1752672 (NO), DMS-1815832 (SP), DMS-1716987 and DMS-1817156 (MV). This work was also partially supported by the grants NSF DMS-1800443/1764366 and the Southeast Center for Mathematics and Biology, an NSF-Simons Research Center for Mathematics of Complex Biological Systems, under National Science Foundation Grant No. DMS-1764406 and Simons Foundation Grant No. 594594 (NJ). The authors thank Eric Reyes and Margherita Ferrari for helpful discussions, and Robert Stolz and the Chédin lab for their help with R-looper and with obtaining the plasmid sequences from their publications.

References

1. Andrew D Bates and Anthony Maxwell. *DNA Topology* Oxford University Press, 2005. ISBN: 978-0198506553.
2. Frédéric Chédin. "Nascent Connections: R-Loops and Chromatin Patterning". In: *Trends in genetics : TIG* 32.12 (Dec. 2016), pp. 828–838. DOI: 10.1016/j.tig.2016.10.002 URL: https://www.ncbi.nlm.nih.gov/pubmed/27793359.
3. Julio Collado-Vides. "Grammatical model of the regulation of gene expression." In: *Proceedings of the National Academy of Sciences* 89.20 (1992), pp. 9405–9409.
4. Shan Dong and David B Searls. "Gene structure prediction by linguistic methods". In: *Genomics* 23.3 (1994), pp. 540–551.
5. Robin D Dowell and Sean R Eddy. "Evaluation of several lightweight stochastic context-free grammars for RNA secondary structure prediction". In: *BMC bioinformatics* 5.1 (2004), p. 71.
6. P. A. Ginno et al. "R-Loop Formation Is a Distinctive Characteristic of Unmethylated Human CpG Island Promoters". In: *Molecular Cell* 45.6 (2012), pp. 814–825. DOI: 10.1016/j.molcel.2012.01.017
7. A. Grigoriev "Analyzing genomes with cumulative skew diagrams". In: *Nucleic Acids Research* 26.10 (Jan. 1998), pp. 2286–2290. DOI: 10.1093/nar/26.10.2286.
8. John E. Hopcroft and Jeffrey D. Ullman. *Introduction to automata theory languages, and computation* Addison-Wesley Series in Computer Science. Addison-Wesley Publishing Co., Reading, Mass., 1979, pp. x+418.
9. F.-T. Huang et al. "Downstream boundary of chromosomal R-loops at murine switch regions: Implications for the mechanism of class switch recombination". In: *Proceedings of the National Academy of Sciences* 103.13 (2006), pp. 5030–5035. DOI: 10.1073/pnas.0506548103.
10. Julian L. Huppert. "Thermodynamic prediction of RNA–DNA duplex-forming regions in the human genome". In: *Molecular BioSystems* 4.6 (2008), p. 686. DOI: 10.1039/b800354h.
11. Bjarne Knudsen and Jotun Hein. "Pfold: RNA secondary structure prediction using stochastic context-free grammars". In: *Nucleic acids research* 31.13 (2003), pp. 3423–3428.
12. Bjarne Knudsen and Jotun Hein. "RNA secondary structure prediction using stochastic context-free grammars and evolutionary history." In: *Bioinformatics (Oxford, England)* 15.6 (1999), pp. 446–454.
13. Lynda Ratmeyer et al. "Sequence Specific Thermodynamic and Structural Properties for DNA.RNA Duplexes". In: *Biochemistry* 33.17 (1994), pp. 5298–5304. DOI: 10.1021/bi00183a037.
14. RW Roberts and DM Crothers. "Stability and properties of double and triple helices: dramatic effects of RNA or DNA backbone composition". In: *Science* 258.5087 (1992), pp. 1463–1466. ISSN: 0036-8075. DOI: 10.1126/science.1279808. eprint: https://science.sciencemag.org/content/258/5087/1463.full.pdf. URL: https://science.sciencemag.org/content/258/5087/1463.
15. D. Roy and M. R. Lieber. "G Clustering Is Important for the Initiation of Transcription-Induced R-Loops In Vitro, whereas High G Density with- out Clustering Is Sufficient Thereafter". In: *Molecular and Cellular Biology* 29.11 (2009), pp. 3124–3133. DOI: 10.1128/mcb.00139-09

16. D. Roy, K. Yu, and M. R. Lieber. "Mechanism of R-Loop Formation at Immunoglobulin Class Switch Sequences". In: *Molecular and Cellular Biology* 28.1 (2007), pp. 50–60. DOI: 10.1128/mcb.01251-07

17. Yasubumi Sakakibara et al. "Stochastic context-free grammers for tRNA modeling". In: *Nucleic acids research* 22.23 (1994), pp. 5112–5120.

18. Lionel A. Sanz et al. "Prevalent, Dynamic, and Conserved R-Loop Structures Associate with Specific Epigenomic Signatures in Mammals". In: *Molecular Cell* 63.1 (2016), pp. 167–178. ISSN: 1097-2765. DOI: https://doi.org/10.1016/j.molcel.2016.05.032. URL: http://www.sciencedirect.com/science/article/pii/S1097276516301964.

19. Science. *Special issue on signals in RNA*. Vol. 352(6292). AAAS, June 2016.

20. Michael Sipser. *Introduction to the Theory of Computation* Vol. 2. Thomson Course Technology Boston, 2006.

21. Robert Stolz et al. "Interplay between DNA sequence and negative superhelicity drives R-loop structures". In: *Proceedings of the National Academy of Sciences* 116.13 (2019), pp. 6260–6269. ISSN: 0027-8424. DOI: 10.1073/pnas.1819476116. eprint: https://www.pnas.org/content/116/13/6260.full.pdf. URL: https://www.pnas.org/content/116/13/6260.

22. Naoki Sugimoto et al. "Thermodynamic Parameters To Predict Stability of RNA/DNA Hybrid Duplexes". In: *Biochemistry* 34.35 (May 1995), pp. 11211–11216. DOI: 10.1021/bi00035a029.

23. Takaaki Yasuhara et al. "Human Rad52 Promotes XPG-Mediated R-loop Processing to Initiate Transcription-Associated Homologous Recombination Repair". In: *Cell* 175.2 (2018), 558–570.e11. ISSN: 0092-8674. DOI: https://doi.org/10.1016/j.cell.2018.08.056. URL: http://www.sciencedirect.com/science/article/pii/S0092867418311176.

24. Kefei Yu et al. "R-loops at immunoglobulin class switch regions in the chromosomes of stimulated B cells". In: *Nature Immunology* 4.5 (July 2003), pp. 442–451. DOI: 10.1038/ni919.

Secondary Structure Ensemble Analysis via Community Detection

Huijing Du, Margherita Maria Ferrari, Christine Heitsch, Forrest Hurley,
Christine V. Mennicke, Blair D. Sullivan, and Bin Xu

Abstract We explored the extent to which graph algorithms for community
detection can improve the mining of structural information from the predicted
Boltzmann/Gibbs ensemble for the biological objects known as RNA secondary
structures. As described, a new computational pipeline was developed, implemented, and tested against the prior method RNAStructProfiling. Since the new
approach was judged to provide more structural information in 75% of the test cases,
this proof-of-principle analysis supports efforts to improve the data mining of RNA
secondary structure ensembles.

H. Du
Department of Mathematics, University of Nebraska-Lincoln, Lincoln, NE, USA
e-mail: hdu5@unl.edu

M. M. Ferrari
Department of Mathematics and Statistics, University of South Florida, Tampa, FL, USA
e-mail: mmferrari@usf.edu

C. Heitsch (✉)
School of Mathematics, Georgia Institute of Technology, Atlanta, GA, USA
e-mail: heitsch@math.gatech.edu

F. Hurley
North Carolina State University, Raleigh, NC, USA
e-mail: fhurley6@gatech.edu

C. V. Mennicke
Department of Mathematics, North Carolina State University, Raleigh, NC, USA
e-mail: cvmennic@ncsu.edu

B. D. Sullivan
School of Computing, University of Utah, Salt Lake City, UT, USA
e-mail: sullivan@cs.utah.edu

B. Xu
Department of Applied and Computational Mathematics and Statistics, University of Notre
Dame, Notre Dame, IN, USA
e-mail: bxu2@nd.edu; bxu@clarkson.edu

© The Association for Women in Mathematics and the Author(s) 2021   55
R. Segal et al. (eds.), *Using Mathematics to Understand Biological Complexity*,
Association for Women in Mathematics Series 22,
https://doi.org/10.1007/978-3-030-57129-0_4

Keywords RNA secondary structure · Gibbs ensemble · Boltzmann probability · Community detection · Graph clustering · Network analysis

1 Introduction

A fundamental principle of molecular biology is that structural information leads to functional insights. By now, accessing the one-dimensional, i.e. sequence, information is generally straightforward. However, for many types of RNA molecules, the most useful information is obtained through experimental determination of three-dimensional structures. Unfortunately, this is often not feasible for the molecule(s) of interest. The alternative is to try inferring higher-order structural information from sequence data. This approach is sufficiently useful that improving structural predictions remains one of the fundamental challenges in computational molecular biology.

One way that structural predictions can be improved is by extracting more information from the available sequence data. For instance, rather than considering a single possible structure (which has minimum free energy under a standard thermodynamic model) for the sequence of interest, we can sample from the entire Boltzmann/Gibbs ensemble according to the free energy approximation in use. This generates a set of possible low-energy structures whose diversity mitigates some of the limitations of the thermodynamic model. Now, however, the problem becomes extracting the most useful structural information from this set of predictions.

Here, we use a graph algorithms approach to address this problem. As described in Sect. 2, we specifically consider RNA secondary structure, which can be understood as the "two-dimensional" structure and is a very useful intermediate representation between the 1D sequence and 3D molecule. Given a representative set of structures, sampled from the Gibbs ensemble according to their Boltzmann probability, prior results [19] introduced the profiling method to separate the structural signal from thermodynamic noise. This is achieved by partitioning the sampled structures into equivalence classes called profiles, and reporting the most probable ones in a summary profile graph. While this method works well on average, we explored the potential to improve the results using community detection approaches from network analysis.

As described in Sect. 3, the first step is to construct an appropriate graph. Toward this end, we considered four different dissimilarity scores on the sampled structures generated by various weightings of the symmetric set difference. The graph of interest then represents structures as vertices joined by edges when the structures have high enough similarity/low enough dissimilarity. We did not set an arbitrary (dis)similarity threshold, but consider a range of sparsification values to find the most informative. We also considered three different community detection approaches for partitioning the graph into clusters. The choice of dissimilarity score, sparsification threshold, and graph clustering algorithm determine a new computational pipeline for RNA secondary structure ensemble analysis via community detection.

The results of this pipeline are reported in Sect. 4. To begin, we implemented a consistent indexing of structural features for a given sequence, which greatly facilitates comparisons across different samples. Second, we allow the sparsification threshold to be determined by the data by adapting a standard clustering quality assessment, the *Caliński-Harabasz* (CH) index [1], to our purposes. This allows us to determine which similarity measure and community detection algorithm are most informative. Finally, we compare the new partitions produced by this pipeline with those from profiling on 12 of the 15 test sequences from [19]. (The other 3 were used for training this new method.)

Overall, we find that, with a reasonable choice of (dis)similarity measure and graph clustering, our new pipeline recapitulates the original profiling results. Moreover, we find that there are specific instances where we identify improvements in the structural information being extracted from the Boltzmann sample. This suggests that the directions discussed in Sect. 5 may be worth pursuing to yield still more informative cluster analyses.

2 Background

We begin this section by describing an RNA molecule and illustrating how we represent a secondary structure using helices. Then, we review computational methods for predicting RNA secondary structures and for analyzing a Boltzmann sample. Particular attention is given to the profiling method [19] which identifies relevant combinations of base pairs in the ensemble, and highlights relationships among sampled structures. Finally, we discuss some challenges in the profiling method which have motivated our work.

2.1 RNA Secondary Structure

Ribonucleic acid (RNA) plays an essential role in many biological processes such as the translation, transcription, regulation and expression of genes. An RNA molecule consists of four types of nucleotide bases: Adenine (A), Cytosine (C), Guanine (G), and Uracil (U). RNA molecules have a secondary structure determined by the hydrogen bonding of complementary bases; the canonical base pairings are the Watson-Crick ones of A-U/U-A and C-G/G-C as well as the wobble pairing of G-U/U-G [5, 27]. Although non-canonical nucleotide interactions are known to exist in RNA molecules, they are not treated as base pairings under the standard nearest neighbor thermodynamic model (NNTM) used for secondary structure prediction [24, 31].

We label our RNA molecules by indexing their bases from 1 to n, the length of the sequence, and use (i, j) to denote a base pair involving nucleotides in positions i and j with $1 \leq i < j \leq n$. A key assumption of the NNTM is that base pairs are

Fig. 1 A secondary structure, drawn by Forna [13], from the Boltzmann sample for the 5S rRNA sequence from *A. tabira* which has length 120 nucleotides (nt). This structure contains 11 helices: $(1, 118, 9)$, $(11, 109, 2)$, $(14, 107, 3)$, $(17, 100, 5)$, $(23, 94, 3)$, $(35, 90, 2)$, $(38, 88, 3)$, $(42, 84, 3)$, $(46, 81, 3)$, $(51, 76, 5)$, $(57, 70, 4)$

noncrossing, i.e. if both (i, j) and (i', j') are in a secondary structure with $i < i'$, then either $i < j < i' < j'$ or $i < i' < j' < j$ but not $i < i' < j < j'$. The excluded case is called a pseudoknot, and accurate prediction of these type of tertiary interactions remains an open problem in the field.

A run of consecutive base pairs of total length k starting with (i, j), i.e. the set of pairings $\{(i, j), (i + 1, j - 1), \cdots, (i + k - 1, j - k + 1)\}$ where neither $i - 1$ and $j + 1$ nor $i + k$ and $j - k$ pair, is called a *helix* and denoted by (i, j, k). A helix of length 1 is called an isolated base pair. In this work, we represent a secondary structure by the set of helices it contains, including any isolated base pairs. Figure 1 illustrates a possible secondary structure for the 5S ribosomal RNA (rRNA) from the fish *Acheilognathus tabira*. Helices are highlighted in blue and green. The green ones will be revisited in Sect. 2.3.

2.2 Computational Methods and Ensemble Analysis

Computational methods for predicting RNA secondary structures based on a nearest neighbor thermodynamic model (NNTM) date back to the 1980s [25]. The goal is to use equilibrium stability, as approximated by the NNTM, as the crucial characteristic for predicting the native base pairings. At first, the focus was on finding a minimum free energy (MFE) secondary structure [29, 32] for the given sequence. It was recognized, though, that this approach can and should be broadened to consider structures within a given energy range of the minimum, e.g. [28, 30]. In this way, thermodynamic optimization can produce biologically relevant structures which are only close to the minimum in free energy under the NNTM.

In part due to the uncertainty as to which low-energy structure(s) were worth investigating, researchers turned to analyzing the distribution of possible secondary structures. The Boltzmann distribution is used to represent a structure's frequency proportional to its free energy. Specifically, for a secondary structure S, its Boltzmann equilibrium probability is defined by

$$P(S) = \exp(-E(S)/RT)/Z,$$

where $E(S)$ is the free energy of the structure, R is the gas constant, T is the absolute temperature, and Z is the partition function for all admissible secondary structures. Here, the partition function is defined by $Z = \sum_S \exp(-E(S)/RT)$ for S ranging over all possible secondary structures for the given sequence. In 1990, McCaskill presented an algorithm to compute the partition function [15]. Later, Ding and Lawrence introduced an algorithm to draw samples from the ensemble of secondary structures [8], which is the basis of most modern ensemble analysis.

There are numerous approaches to ensemble analysis for understanding secondary structures arising from a Boltzmann sample [2, 6, 7, 9, 12, 19, 22]. Though the number of possible structures can increase exponentially as the sequence length increases [26], a Boltzmann sample of 1000 structures is typically used for mining representative structures [8]. In [6, 7], Ding, Chan, and Lawrence used a clustering method to group structures. Their method represents a secondary structure by a symmetric 0-1 matrix A whose entry A_{ij} is 1 if and only if the base i pairs with the base j. The distance between two structures is measured by the number of different base pairs. Using a hierarchical clustering method, they computed clusters and represented each cluster by the centroid. Another way to group structures is to use the branching configuration of a structure; Giegerich, Voß, and Rehmsmeier introduced the concept of RNA shape and presented an algorithm to compute the shapes [9]. Structures with the same shape belong to an abstract shape class, which is represented by its MFE member. Finally, profiling [19] is a combinatorial approach that identifies the dominant combinations of base pairs in the ensemble, and uses them to highlight similarities and differences across sampled structures. We describe this approach in further detail in the next section.

2.3 Profiling

In this section, we will describe the profiling pipeline (illustrated in Fig. 2) and discuss some current challenges of this method. Profiling takes a Boltzmann sample as input, where the secondary structures are represented as sets of helices.

Helix classes and features In the first step, profiling partitions the set of all observed helices into "helix classes" using maximal helices and chooses the highest frequency ones as "features." A helix $g = (i, j, k)$ is said to be *maximal* if neither $(i - 1, j + 1)$ nor $(i + k, j - k)$ would be canonical base pairs. If $(i - 1, j + 1)$ is

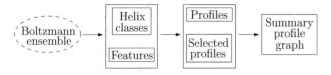

Fig. 2 A schematic representation of the profiling pipeline

not a canonical base pair, but $(i + k, j - k)$ is with $j - i - 2k < 4$, then the helix g is also maximal. These conditions insure that a maximal helix is non-extendable using the canonical Watson-Crick or G,U wobble pairings and does not end with an hairpin of length less than 3.

Every maximal helix g determines a *helix class* $c_g = [g] = [(i, j, k)]$, which is the equivalence class of all helices h which are subsets of the set of pairings $\{(i, j), \ldots, (i + k - 1, j - k + 1)\}$. Since the frequency of a helix h is the number of times h appears in the sample, the frequency of $[g]$ is defined as the sum of the frequencies of its elements. In the following, we will be dealing exclusively with helix classes, and not individual helices, so will drop the square bracket equivalence class notation. When considering a specific RNA sequence, helix classes will be assigned a unique nonnegative integer index as described in Sect. 4.1. When discussing a particular Boltzmann sample, we will either refer to the helix class by this integer index and/or by its (i, j, k) triple (without the square brackets).

To identify the most "important" runs of base pairs, called *features*, helix classes are ordered by decreasing frequency, and selected up to some threshold. By default, profiling sets the threshold to be the point of diminishing returns in the frequency distribution, computed using Shannon entropy [21]. The table in Fig. 3 lists the features for the 5S rRNA from *A. tabira* with the indices described in Sect. 4.1. Three of the four most frequent features appear in Fig. 1 and are the helices colored green. As it happens, all of the green helices are maximal. However, the helix (11, 109, 2) is not, since it could be extended by the (10, 110) canonical C-G pairing.

Profiles and selected profiles In the second step, profiling consolidates similar secondary structures according to their *profiles*, which consist of the set of features present (from step one). Hence, a profile p is an equivalence class of structures having the same set of features, although their exact base pairings may vary somewhat since the features are themselves equivalence classes. Additionally, a profile says nothing about base pairings from helix classes which are not featured. The secondary structure in Fig. 1 has only three features (in green) which are indexed 0, 28, and 17, cf. Fig. 3. The profile for this structure is denoted by the string '[0[28[17]]]' rather than {0, 17, 28}; the square brackets indicate the nesting of features, i.e. their relative positioning in the secondary structures,

The number of secondary structures in the Boltzmann sample with a particular profile is called its *specific frequency*. Since profiling seeks to highlight relationships among the sampled structures, it is also useful to define the *general frequency* of a profile p to be the number of structures sampled whose profile is a superset of p. Similarly to feature selection, this stage of the pipeline concludes by again downselecting by thresholding the specific frequency distribution, producing *selected profiles*, which can be used to compare and contrast structures in a summary profile graph as shown in Fig. 3.

Summary profile graph Profiling outputs a summary profile graph illustrating relationships between selected profiles. It is inspired by the Hasse diagram for the subset relation, although it is 'read' from smallest/top to largest/bottom sets. The

Fig. 3 Summary profile graph for the 5S rRNA from *A. tabira*. The secondary structure for this sequence illustrated in Fig. 1 has features 0, 17, and 28 highlighted in green. Since that structure has no other base pairings which belong to a featured helix class, it is one of the 43 structures sampled whose profile is [0[28[17]]]. However, the general frequency of this selected profile is much higher, since 809 of 1000 structures sampled include these three features

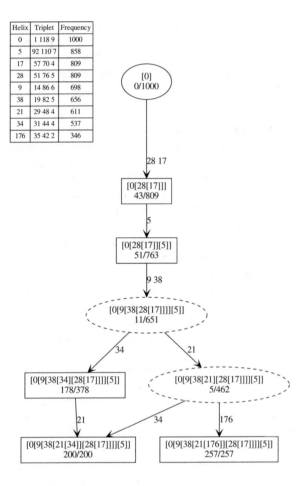

Helix	Triplet	Frequency
0	1 118 9	1000
5	92 110 7	858
17	57 70 4	809
28	51 76 5	809
9	14 86 6	698
38	19 82 5	656
21	29 48 4	611
34	31 44 4	537
176	35 42 2	346

goal is to facilitate comparisons in this graph among the high frequency structural classes identified by profiling the Boltzmann sample.

The vertex set consists of the selected profiles as well as all nonempty intersection profiles, with the possible addition of a 'root' profile. The intersection profiles 'interpolate' between selected profiles which are not comparable under the subset relation. More precisely, consider a profile p as a set of features. Then if $P = \{p_1, \ldots, p_s\}$ are the selected profiles, the *intersection profiles* are $I \setminus (P \cup \emptyset)$ where $I = \{p_i \cap p_j \mid 1 \leq i < j \leq s\}$. These additional vertices make explicit the relationship among the selected profiles. The graph is rooted at the profile common to all sampled structures, which may be the empty set in this case.

In the graph, selected profiles have a solid, rectangular outline while intersection ones have a dashed oval to clearly distinguish the two types. Each vertex is labeled with the profile string along with the ratio between the specific and general frequencies. Note that while the specific frequency of an intersection profile $p_i \cap p_j$ may be 0, the general frequency will be at least the sum of the specific frequencies of both p_i and p_j. If the root profile is neither a selected nor an intersection profile, then it is drawn as a solid oval. Its specific frequency may be 0, but by definition the general frequency is the sample size (typically 1000 structures).

Two vertices p and p' in the graph are connected by an edge, understood as from p to p', if $p \subsetneq p'$ and there is no $p'' \in P \cup I$ with $p \subsetneq p'' \subsetneq p'$. The edge is labeled with the feature(s) from $p' \setminus p$.

Challenges in profiling Profiling captures the dominant structural patterns appearing in the Boltzmann sample, and highlights similarities and differences among structures using the summary profile graph. The method provides informative, detailed, and reproducible results [19], and may suggest different conformations for regions of interest, as shown in the proof-of-principle analysis. However, while generally useful, there are ways in which the profiling approach might be improved. The example in Fig. 3 illustrates two of them.

Profiling is explicit about separating structural signal from thermodynamic noise. As a consequence, it is not unusual to have sampled structures which do not appear in any selected profile. For example, we see from the summary profile graph in Fig. 3 that there were almost 200 structures sampled for the *A. tabira* 5S rRNA sequence that do not contain features 28 and 17. Since this frequency is below the threshold for feature selection, we do not know if all of those structures contain a common alternative. Retaining information about alternatives to high frequency features is one way that profiling could be enhanced. This could be particularly useful when seeking to understand RNA molecules whose functionality may depend on switching from one conformation to another.

When considering the list of features for *A. tabira*, we see that the maximal helix 34 overlaps with 21 and with 176, i.e. there is a least one sequence index which appears in both feature's defining set of base pairs. However, since the sampled base pairs can be a subset of the maximal helix, features 34 and 21 can – and do – coexist in a secondary structure, and appear together in a selected profile. In contrast, the overlap between 34 and 176 is substantial enough that they do not appear together in a selected profile. Recognizing unions of features which are consistently sampled would enhance profiling by consolidating their frequencies into a stronger structural signal and reducing the number of selected profiles as well as the complexity of the summary profile graph. This should better highlight the structural signal present in the Boltzmann ensemble.

Finally, the complexity of the summary profile graph is also a concern which has been identified in some of the profiling outputs, although not apparent in the *A. tabira* example. This often seems to be the result of a combinatorial explosion in the number of intersection profiles, perhaps as a consequence of overlapping features which should be treated as a union, leading to an incomprehensible graph. These types of challenges motivated us to consider alternative approaches to analyzing the secondary structures in a Boltzmann sample.

3   Methods

As described in Sect. 2.3, profiling focuses on denoising the distribution of helix classes by identifying as features those with high enough probability, and restricting attention to these features when generating a profile of each structure. However, potential structural alternatives are not captured due to thresholding the frequency distribution. To capture the structural alternatives in a comprehensive manner, we opt to represent the secondary structures by their entire set of helix classes (not just the featured helix classes), which we refer to as the *extended profiles*. Here we describe the methods used to compare structures using these extended profiles. First, we introduce similarity measures calculated by comparison of the sets of helix classes, then we describe graph clustering algorithms used for grouping together structures based on these similarities.

3.1   Similarity Measures

Although we refer to these computations as our similarity measures, we will actually be computing the dissimilarity between extended profiles. To quantify the (dis)similarity among structures, we introduce several distance measures based on set symmetric difference. Using extended profiles, each structure is represented as a set of helix classes (typically denoted c_i), and a natural notion of difference arises from considering how much these sets differ. Since not all helix classes are equally important structurally (e.g. some may represent much higher frequency structures or much lower energy configurations), we consider several weightings of the symmetric difference in our analysis. The symmetric difference, or disjunctive union, of two sets A and B is denoted $A \triangle B$ and is the set of elements which are in their union $A \cup B$ but not their intersection $A \cap B$.

Base similarity measure We define our *base similarity measure* of two structures S and T to be the number of helix classes they disagree on, or more formally, the cardinality of the symmetric difference of their extended profiles:

$$m_{\text{none}}(S, T) = |\{c_i \in S \triangle T\}|.$$

For example, consider the structures $s_1 = \{c_1, c_2, c_3\}$, $s_2 = \{c_1, c_2, c_4, c_5\}$ on helix classes c_i, $1 \leq i \leq 5$. The base similarity measure of these two structures is then $m_{\text{none}}(s_1, s_2) = |(s_1 \cup s_2) \setminus (s_1 \cap s_2)| = |\{c_3, c_4, c_5\}| = 3$.

Length-weighted measure To incorporate the idea that two structures which differ only on a short helix class are closer than two which differ on inclusion of a very long helix class (measured by number of base pairs), we introduce a *length-weighted similarity measure*. Using the standard notation that a helix class $c_n = (i_n, j_n, k_n)$

has start position i_n, end position j_n, and length k_n, the length-weighted similarity measure is

$$m_{\text{length}}(S, T) = \sum_{c_i \in S \Delta T} k_i.$$

Using the same example given for the base measure, $m_{\text{length}}(s_1, s_2) = k_3 + k_4 + k_5$.

Frequency-weighted measure Alternatively, it is plausible that frequency in the Boltzmann ensemble is a better proxy for helix class importance than length, and thus we introduce a measure where the symmetric difference is weighted by f_i, the frequency of helix class c_i:

$$m_{\text{frequency}}(S, T) = \sum_{c_i \in S \Delta T} f_i.$$

Here, f_i is a proxy for the probability of the helix class c_i (which is the sum over all the frequencies of its helices). In our running example, $m_{\text{frequency}}(s_1, s_2) = f_3 + f_4 + f_5$.

Energy-weighted measure Finally, we incorporate an approximation of the free energy e_n for a helix class c_n as the weighting function in our symmetric difference:

$$m_{\text{energy}}(S, T) = \sum_{c_n \in S \Delta T} e_n.$$

sHere, e_n is the sum of the free energy associated with each base pair in the helix in c_n. For each helix, we compute a weighted sum of the number of Watson-Crick pairings A$\leftrightarrow$U and C$\leftrightarrow$G as well as wobble pairings G$\leftrightarrow$U, using weight 2 for a A$\leftrightarrow$U pair, 3 for a C$\leftrightarrow$G pair, and 1 for a G$\leftrightarrow$U pair to approximate the energy, similar to Nussinov's algorithm [16]. In our example, $m_{\text{energy}}(s_1, s_2) = e_3 + e_4 + e_5$, where $e_n = 2l_{\{n,AU\}} + 3l_{\{n,CG\}} + l_{\{n,GU\}}$, and $l_{\{n,\alpha\beta\}}$ represents the total number of $\alpha \leftrightarrow \beta$ pairs in a helix class c_n.

3.2 Graph Clustering

A common task in graph analytics is to identify communities (clusters) of "similar" or "well-connected" vertices based on the structure of edges in the network; the methods used to do this are called *graph clustering algorithms*. Typically, communities are disjoint (though some algorithms exist for detecting overlapping communities), and methods try to optimize measures that formalize the notion that nodes within a community should be better connected to each other than to nodes in other communities. Where available, many algorithms can incorporate weights on the edges (e.g. indicating strength of similarity or association). The quality of a partition is evaluated based on measures such as low variance (or high density) within clusters and high separation (low edge density) between clusters, and some tuning of the algorithm may be required to produce an optimal clustering. We refer the reader to [20] for an overview of common methods.

To utilize graph clustering on the set of extended profiles, we form a graph $G = (V, E)$, where V is the set of extended profiles from a Boltzmann sample, each labeled with its frequency of occurrence in the sample. We note that the frequencies will sum to the sample size (and $|V|$ will typically be smaller due to duplicate structures and/or similar structures in the same equivalence class). The set E will initially contain an edge between every pair of vertices, weighted using a similarity score as defined in Sect. 3.1.

Prior to clustering, we sparsify our input graph by retaining only edges between the most similar structures. (Recall that the measures from Sect. 3.1 actually calculate *dis*similarity.) This is done by setting a similarity threshold corresponding to a fixed percentile of the edge weight distribution (e.g. we keep the bottom 40% of edges) to ensure equity across samples. This step enables the application of (unweighted) graph community detection algorithms by removing connections between less similar vertices. In general, deleting more edges creates sparsity and generally increases the number of clusters formed. We discuss threshold selection in Sect. 4.2. Figure 4 shows an example of thresholding a synthetic instance to remove 70% of the edges.

We tested several graph clustering algorithms using the implementations in NetworkX [11], and now describe each method individually.

Greedy Modularity Community (GMC): Also known as Clauset-Newman-Moore, GMC is a hierarchical clustering method that calculates modularity to determine an optimal graph partition. A graph's modularity is high when edge connections are dense within communities and sparse between communities; an exact formula for calculating modularity is given in [3]. Thus, a "good" partition

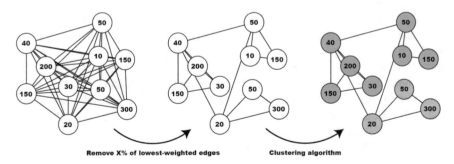

Fig. 4 Community detection in the set of extended profiles. The initial input is a graph $G = (V, E)$ with vertices V representing extended profiles labeled with their frequency in the Boltzmann sample. Edges E are weighted with the (dis)similarity score, and a percentage of edges are removed prior to clustering. The clustering algorithm then groups vertices using the remaining edge information

should have a high modularity. To use this property for clustering, GMC begins with each vertex in its own isolated community. The algorithm then iteratively merges pairs of communities by determining which pair's union produces the largest increase in the graph's modularity, stopping when no such pair exists. That is, the algorithm finds the partition of the graph with the maximal modularity.

Girvan Newman (GN): In contrast with GMC, the GN algorithm begins with all vertices in the same community and progressively removes edges at each iteration of the algorithm [10]. An edge is selected for removal if it has the highest betweenness centrality, defined as the number of shortest paths between pairs of vertices that include that edge. The algorithm continues until all edges have been removed, or until a specified number of iterations have been completed.

Label Propagation Algorithm (LPA) In this algorithm, each vertex begins with a unique label, and labels are updated iteratively to match the most frequent label among neighbors [4]. Ties are broken randomly in the case where multiple labels are equally frequent among a vertex's neighbors. This nondeterministic step can produce different clusterings from multiple runs of the algorithm, which we discuss in Sect. 4.3.

The output of graph clustering is a partition of the set of extended profiles. The number of communities generated depends on the clustering algorithm used and the sparsification threshold; we discuss choosing the optimal threshold using the CH index for cluster evaluations in Sect. 4.2. For easy interpretation, we also visualize the clustering by color-coding the nodes in an output graph, shown in Fig. 4.

3.3 Pipeline

The overall flow of the process is shown in Fig. 5. We used **RNAstructure**[1] from the Mathews Lab at the University of Rochester to perform the initial sampling of secondary structures from the Boltzmann distribution. After minor data cleaning and application of consistent indexing to allow comparison across samples, we then calculate pairwise similarity between structures, construct similarity graphs, and perform graph clustering. All internal code is written in Python 3 and available under a BSD 3-clause license at **Github.com**.[2] We compared the clustering results to the outcome of profiling as implemented in **RNAStructProfiling**.[3]

Based on experimental results (see Sect. 4.3), we constructed similarity graphs using the frequency metric described in Sect. 3.1 and applied the GMC clustering algorithms described in Sect. 3.2. Clustering quality was assessed using a variant of the standard CH index, described in Sect. 4.2.

---

[1] https://rna.urmc.rochester.edu/RNAstructure.html

[2] https://github.com/gtDMMB/ipam-wbio-scripts

[3] The results in this manuscript use a new version of profiling which is still under development and will be made available at https://github.com/gtDMMB/RNAStructProfiling.

System Diagram

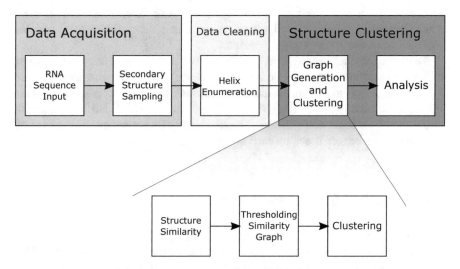

Fig. 5 The high level components in the RNA structure clustering pipeline

4 Results and Discussion

As illustrated in Fig. 5, the new structure analysis pipeline depends on three different components: the similarity score, the thresholding percentage, and the clustering algorithm. Below we will describe how each of these was determined from the options explained above, but first we report an improvement to the code base (consistent indexing) which greatly facilitated comparisons between the original profiling partitions and the new ones produced by graph clustering. Next, we describe how the thresholding was handled, since different levels of graph sparsity were found to generate significantly different partitions. We then address the choice of similarity score, and then the choice of clustering algorithm. Finally, we report the results of our new analysis pipeline when tested against 12 of the original 15 test sequences [19] from four different RNA families. We note that the three transfer RNA (tRNA) sequences from the original test set were not considered, since they had been used extensively for training when developing this new approach.

4.1 Consistent Indexing

As originally implemented, profiling indexed helix classes according to their frequency in the Boltzmann sample currently being analyzed. Since the sampling is stochastic, these indices can vary somewhat between different samples (e.g.

(10, 17, 3) might be helix class 1 in sample S_1 but helix class 3 in sample S_2). To improve analysis of secondary structures, we introduced a consistent labeling system so that a given helix class is assigned the same label in any sample for a given sequence. We begin by assigning a canonical complete ordering to all possible maximal helix classes for a given sequence.

To enumerate maximal helices, we iterate over all possible start and end points of helices (starting from either end of the sequence) and check whether they were matched in a previously identified helix class. If it is a new pairing we then iteratively increase the helix length until it becomes maximal and add the result to our list of maximal helix classes. The maximal helix classes are then sorted by length with ties broken based on earliest start location.

4.2 Clustering Quality Assessment

The partition generated by our new pipeline depends on the similarity measure, threshold percentage, and clustering algorithm used. To select the most useful of the four similarity measures and three clustering algorithms considered, we first address how to determine a threshold percentage for the graph sparsification. This will be done by adapting the standard *Caliński-Harabasz* (CH) index [1] for our purposes.

The CH index is often used to determine an optimal number of clusters. It is a "variance ratio criterion" [1] intended to capture the intuition that, ideally, points within a cluster are "close together" while the clusters themselves are "far apart." Let $\{C_1, \ldots, C_k\}$ denote a partition of a data set, such as an ensemble of structures, of size n into k clusters. The CH index is the ratio of the between-cluster sum of squares $B(k)$ to the within-cluster sum of squares $W(k)$;

$$CH(k) = \frac{B(k)/(k-1)}{W(k)/(n-k)}.$$

The numerator measures the sample variance between the different groups, computed based on the distances between the centroid of the n data points and the centroids of each cluster, while the denominator averages it over each of the k clusters. More precisely,

$$B(k) = \sum_{i=1}^{k} n_i d^2(C_e^*, C_i^*) \quad \text{and} \quad W(k) = \sum_{i=1}^{k} \sum_{S \in C_i} d^2(C_i^*, S),$$

where $d(C_e^*, C_i^*)$ is the distance between the ensemble centroid C_e^* and the centroid C_i^* of the i-th cluster which contains n_i data points, while $d(C_i^*, S)$ is the distance between the centroid C_i^* and a structure S in the i-th cluster. Following [6], the centroid of the ith cluster C_i^* is the secondary structure realizing the minimum total distance to all the other structures in C_i. The ensemble centroid C_e^* is defined similarly by considering all structures sampled. Finally, to simplify the calculation slightly, the distances were not squared.

Typically, the clustering with the maximum CH index is taken to be the best one. However, we found that too often the maximum CH index was obtained for the minimum graph sparsity, and the resulting partition was evaluated as being too fine grained. Instead, we identified the maximum proportional increase in CH index as a characteristic which correlates with "good" partitions. More specifically, we focus on the clustering with the highest value of

$$\frac{CH_{i+1}}{CH_i} \quad \text{for } i = 1, 2, \ldots. \tag{1}$$

where CH_i is the CH index obtained for the i-th threshold considered.

This approach was validated on the 5S rRNA sequence from the archaea *Desulfurococcus mobilis*, as summarized in Table 1 and Fig. 6. The maximum CH index is reached at edge sparsification threshold 99, and corresponds to a partition having 28 clusters. Since there are 34 distinct extended profiles, this partition places the majority of structures into clusters of size 1, a highly granular approach that fails to reveal many significant similarities between structures. However, the biggest proportional increase in CH index values occurs at threshold 96 for a partition with 22 clusters. By direct inspection, we confirmed that each selected profile gets its own cluster having mass equal to the specific frequency of the corresponding profile, meaning that this clustering is providing the same information as profiling. Importantly, we also found that such a partition was identical to our carefully hand-crafted "ground truth" partition of the extended profiles.

A similar behavior was observed for other test sequences, i.e. a monotone increasing CH index with the sparsification of the similarity graph but a significant proportional increase at some critical edge sparsification threshold. Hence, we used

Table 1 CH index evaluation of the clustering for the *D. mobilis* 5S rRNA. The similarity measure used was the frequency-weighted symmetric difference, and the clustering was computed with the GMC graph algorithm. The quality of the clustering was manually evaluated against a hand-crafted "ground truth" partition of the extended profiles

Threshold	# of Clusters #	CH index	Proportional increase
10–60	1	N/A	N/A
70	3	5.052746249119672	N/A
80	5	31.676599148009707	~6.27
85	8	303.22637579865153	~9.57
90–92	9	265.8620083317187	~0.88
93–95	14	1507.975920763893	~5.67
96–97	22	164354.62111801244	~108.99
98	27	337551.99572649575	~2.05
99	28	471345.67741935485	~1.40

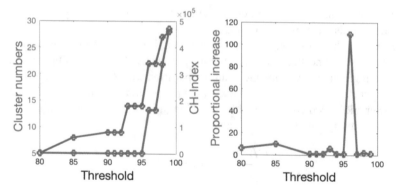

Fig. 6 CH index value, number of clusters, and proportional increase of CH index values for sequence *D. mobilis*

the largest proportional increase in the CH index to evaluate the quality of the similarity measure and clustering algorithm in our analysis pipeline.

4.3 Similarity Measure and Clustering Algorithm Selection

To determine which similarity measures and community detection methods were most suitable in the clustering pipeline, we initially began with the four choices of similarity measure defined in Sect. 3.1 and three choices of community detection algorithm listed in Sect. 3.2. We then clustered a selection of RNA sequences over a range of edge thresholds using each pair of similarity measure/community detection algorithm in the clustering pipeline. The pair of methods that produced the best quality of clusterings was selected.

Besides the previously described maximum proportional CH index, several criteria were used in assessing the quality of the clusterings produced with different paired methods. First, we desired consistency, as measured by the method's ability to reproduce a particular clustering over multiple runs of a sequence through the pipeline. Second, we wanted a method that produced a monotonic increase in the total number of clusters with increasing edge thresholds, preferably without large jumps in the number of clusters so that we could evaluate a more gradual refinement of the partition. Finally, an optimal method would separate the selected profiles from profiling into different clusters, allowing for more direct comparison between the clustering and profiling results. From assessing the clusterings formed from different pairs of similarity measures and community detection algorithms, we concluded that the combination of $m_{\text{frequency}}$ and GMC overall produced the best clustering results across all sequences tested. We summarize this assessment below.

Choice of similarity measure Not surprisingly, m_{none} did not perform as well as the weightings incorporating additional biological information. For instance, on one of the test sequences, the number of clusters jumped from 2 to 18 when increasing the edge threshold under m_{none}, whereas $m_{frequency}$ produced a more gradual increase in total clusters. Thus, m_{none} was quickly eliminated as a candidate and we further examined m_{length}, m_{energy}, and $m_{frequency}$ on additional sequences.

For *D. mobilis*, m_{length} and m_{energy} produced clusters in which selected profiles from profiling were placed in the same cluster, while $m_{frequency}$ separated them. Additionally, we noticed that the CH index for $m_{frequency}$ clusterings was higher than those formed using m_{length} and m_{energy} at the same threshold. Both of these pointed to $m_{frequency}$ as a better option for clustering.

It is worth noting that in a Boltzmann distribution, the probability of a structure's occurrence (and hence its frequency in a Boltzmann sample) is inversely proportional to the exponential of its energy. Consequentially, when structure similarities were weighted using m_{energy}, edge removal tended to favor separating high energy, low probability structures while keeping the higher probability structures in a single cluster. Edge removal thresholds $>90\%$ were generally required to separate the high probability structures into separate clusters. Thus, while m_{energy} and $m_{frequency}$ are related weights, the latter requires lower thresholds to form distinct clusters separating the high probability structures, and it can do so with greater refinement.

Lastly, we found that $m_{frequency}$ was more easily interpretable in terms of comparing to profiling results, since profiling already takes into account the frequency of the helix classes. Using $m_{frequency}$ for clustering may therefore aid in building on the profiling results, as it may be possible to perfectly reproduce the selected profiles in clusters generated using frequency for some threshold.

Choice of community detection algorithm In evaluating the community detection algorithms, we not only examined sample sequence clusterings, but also compared the algorithms' mechanics. First, we found that under the LPA algorithm, there was significant variation in the partitions produced from multiple runs on the same sample for a given sequence. This can be explained due to the fact that LPA has a non-deterministic step in its algorithm as described in Sect. 3.2, but we chose to focus on algorithms with reproducible clusterings in this initial study.

Generally, GMC and GN produced similar results at a given edge sparsification threshold, and performed comparably on our sample sequences. GMC inherently optimizes the modularity which in turn determines the number of clusters. On the other hand, GN requires setting a number of iterations to output a fixed level of the hierarchical clustering. While we acknowledge that this could be automated using an assessment of the clustering quality at different hierarchical levels (using e.g. the CH index), we felt that increasing our pipeline's reliance on such a metric was less desirable, and thus chose to focus on GMC for our experiments.

4.4 Test Sequences

In the next section, we analyze the results of the clustering pipeline, and compare it with the output of profiling. To assess what improvement this new approach could achieve, we compared the community detection partitions with the profiling results for 12 of the original 15 test sequences [19].

As will be described, we considered three sequences from each of four different RNA families: quorum regulatory RNA (Qrr), 5S ribosomal RNA (rRNA), tetrahydrofolate (THF) riboswitch and thiamine pyrophosphate (TPP) riboswitch. (The three from the tRNA family were used for training the method.) An RNA family is a set of homologous sequences which have been found to have a similar secondary structure, and hence a common functionality across different organisms. The canonical example is, of course, the tRNA family.

Small non-coding RNA molecules called Qrr have been shown to regulate the expression of virulence genes in bacterial pathogens [17]. We examine the results of the clustering pipeline in comparison with the original profiling results for three Qrr sequences found in members of *Vibrionaceae*, a family of aquatic bacterial pathogens: Qrr1 and Qrr3 from *Vibrio cholerae* and Qrr1 from *Vibrio harveyi*. *V. cholerae* is the human pathogen responsible for causing cholera; it contains four Qrr molecules that can each affect translation of virulence genes. The common aquaculture pathogen *Vibrio harveyi* contains five Qrr molecules (Qrr1-5) that all must be present to regulate gene expression [17].

THF serves as a coenzyme for metabolic reactions. In bacteria, regulation of THF biosynthesis and transport is associated with THF riboswitches. THF riboswitches are a class of RNAs that can bind to THF. These riboswitches are found in Firmicutes which include *Clostridium botulinu*, *Streptococcus uberis*, and *Mitsuokella multacida*.

The TPP riboswitch class is the most widespread riboswitch occurring in all domains of life. Even though it controls different genes involved in the synthesis or transport of thiamine and its phosphorylated derivatives in bacteria, archaea, fungi, and plants, TPP is a highly conserved RNA secondary structure. We analyze three TPP sequences: *Bacillus clausii*, *Pasteurella multocid*, and *Thermoplasma acidophilum*. *B. clausii* is an alkaliphilic bacterium used for production of a high-alkaline protease (M-protease) in laundry detergents. *P. multocid* is the cause of a range of diseases in mammals, birds, and humans. *T. acidophilum* is an archaeon, originally isolated from a self-heating coal refuse pile, at pH 2 and 59 °C.

5S rRNA is a structural and functional component of the ribosome. Although it is found in all organisms, with the exception of mitochondrial ribosomes of fungi and animals, its exact function is not fully understood. The secondary structure of a 5S rRNA is generally described by a model obtained from sequence and structural analysis. We examine the results of the clustering pipeline for the 5S rRNAs *Desulfurococcus mobilis*, *Acheilognathus tabira* and *Escherichia coli*, highlighting differences from those obtained by profiling. *D. mobilis* is an extreme thermophile that lives at temperatures of up to 97 °C in solfataric hot springs, where the pH is between 2.2 and 6.5. *A. tabira* is a species of ray-finned fish that is endemic in Japan and has 5 subspecies. *E. coli* is an anaerobic bacterium that normally lives in the intestines of warm-blooded organisms. Most varieties are harmless, but some strains can cause diseases.

4.5 Results of Experiments

The new analysis pipeline was tested on 12 sequences, 3 each from 4 different RNA families, as described in Sect. 4.4. Results were compared against those from profiling [19], and are summarized in Table 2. To illustrate the methodology, we will consider two sequences as examples. The complete data for all sequences — original Boltzmann sample, summary profile graph, clustering partition file, and visualization of the clustering graph—are available in individual directories at https://github.com/gtDMMB/ipam-wbio-scripts/Data.

In Table 2, the NCBI accession numbers are given when available, and the sequence number with publication reference otherwise. No correlation was observed between the results and sequence length (Len).

The difficulty (Diff) of the profiling analysis for each test sequence was classified as easy (E), medium (M), or hard (H) as described next. Two factors were considered in this assessment: the number of selected profiles (SP) and the Boltzmann sample coverage (Cov). For completeness, we also report the number of features (Feat) identified and the number of profiles (Prof).

Recall that a profile is an equivalence class of sampled structures determined by a maximal combination of features. Selected profiles are chosen by an information entropy thresholding criteria. Coverage is the percentage of the Boltzmann sample belonging to one of these selected profiles. A sequence was classified as *easy* if it had 3–5 selected profiles (and hence a relatively simple profile graph), *and* coverage of at least 75% of the sampled structures. If only one of these conditions was met, then the sequence was classified as *medium* difficulty, while if both failed, then it was *hard* to analyze.

The new analysis pipeline is based on extended profiles, which are the equivalence classes of secondary structures determined by all the helix classes in a sample, not just the features. In general, the number of extended profiles (EP) is much higher than the number of profiles. Threshold (Thresh) is the sparsification cutoff, i.e. the percentage of edges that were removed from the complete graph according to the edge weight distribution determined by the (dis)similarity measure. The optimal threshold was determined using a variant of the CH index as described in Sect. 4.2. Using the GMC community detection algorithm on the sparsified graph results in a number of clusters (Clust).

For a given sequence, we analyzed the clustering partition realizing the biggest proportional increase in the CH index. The pipeline produces a detailed partition file that lists all extended profiles in each cluster along with their frequencies (see Figs. 9 and 10 in Sect. 5 for examples). A visualization of the graph clustering was also

Table 2 Summary of the sequences tested, profiling analysis, and results from new community detection pipeline

Family	Organism	Accession	Len	Diff	SP	Cov	Feat	Prof	EP	Thresh	Clust	Recap	Imprv
Qrr	*V. cholerae*	#1,[14]	96	E	3	96.5	7	9	46	48	3	No	No
Qrr	*V. cholerae*	#3,[14]	107	E	4	98.4	6	8	66	67	7	No	Yes
Qrr	*V. harveyi*	#1,[23]	95	M	2	70.7	8	19	74	53	3	No	Yes
THF	*C. botulinum*	CP000939	101	E	5	76.0	9	46	143	96	45	Yes	No
THF	*S. uberis*	AM946015	91	M	5	60.5	11	81	141	99	85	Yes	No
THF	*M. multacida*	ABWK02000009	99	H	9	62.8	12	69	308	97	51	No	Yes
TPP	*B. clausii*	AP006627	100	E	5	78.3	10	41	69	87	13	No	Yes
TPP	*P. multocid*	AE004439	93	M	13	79.9	11	54	146	96	34	No	Yes
TPP	*T. acidophilum*	AL445064	107	H	7	71.8	7	35	271	99	120	No	Yes
5S	*D. mobilis*	X07545	133	E	4	82.4	11	22	34	96	22	Yes	Yes
5S	*A. tabira*	AB015591	120	M	5	72.9	9	25	216	94	24	Yes	Yes
5S	*E. coli*	V00336	120	H	13	70.4	11	80	249	97	38	No	Yes

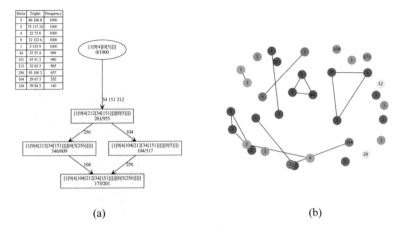

(a)   (b)

Fig. 7 Clustering for *D. mobilis* 5S rRNA both recapitulates and improves on profiling because, in addition to capturing the same strong signal present in the ensemble, it retains information about feature 154 (which was not part of any selected profile). See Fig. 9 on page 77 for details on the extended profiles in each cluster. (**a**) *D. mobilis* summary profile graph (**b**) *D. mobilis* clustering (threshold 96). Each node represents an extended profile, labelled with frequency and colored by cluster

generated for each sequence (see Figs. 7b and 8b for examples), but was found to be of minimal assistance in interpreting the results. The profiling and graph clustering analyses were compared (by painstaking hand comparison of the extended profiles) for each of the sequences to determine if the graph clustering *recapitulates* (Recap) the profiling results, which was generally desired, and/or if it *improves* (Imprv) on them by adding new information not observed from profiling.

A clustering partition was judged to recapitulate profiling (Recap) if the selected profiles were separated into distinct clusters. For instance, the clustering shown in Fig. 7b for the *D. mobilis* 5S rRNA (see Fig. 9 on page 77 for details) has the 4 largest clusters corresponding to the selected profiles in Fig. 7a. However, only a third of our test sequences satisfied this condition, as illustrated by the second example. In contrast, the clustering for *V. cholerae* Qrr3 shown in Fig. 8a does not recapitulate profiling; two of the selected profiles shown in Fig. 8a are grouped in a single cluster (profiles [[1][13[20]][10][0]] and [[1][13[20]][0]] in Fig. 8a, red cluster in Fig. 8b, and Summary 0 in Fig. 10 on page 78). For this sequence, these two selected profiles were not separated clusters until a higher edge sparsification threshold. Generally, we observed that while there was some clustering partition for each sequence which did distinguish selected profiles, this was not necessarily true at the threshold selected using the CH index criteria.

When evaluating whether a clustering partition improved on profiling, we considered four criteria. First, was the percentage of the sample covered by the clusters which contained one or more selected profiles higher than the profiling coverage? Second, did clustering recover non-featured helix classes which were common to all structures represented by some selected profile? Third, did clustering

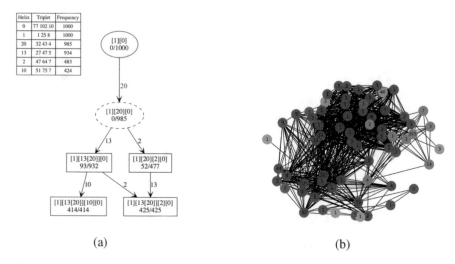

Helix	Triplet	Frequency
0	77 102 10	1000
1	1 25 8	1000
20	32 43 4	985
13	27 47 5	934
2	47 64 7	483
10	51 75 7	424

(a)   (b)

Fig. 8 Clustering for *V. cholerae* Qrr3 sRNA does not recapitulate profiling (it combines [[1][13[20]][10][0]] and [[1][13[20]][0]] into a single cluster), but is still an improvement. Specifically, all profiles in this new cluster missing helix class 10 alternatively contained less frequent helices over the same general region of base pairs; this implies it is more informative to consider the two selected profiles as part of one larger group of structures. See Fig. 10 on page 78 for details on the extended profiles in each cluster. (a) *V. cholerae* Qrr3 summary profile graph (b) *V. cholerae* Qrr3 clustering (threshold 67). Each node represents an extended profile, labelled with frequency and colored by cluster

recover structures containing a feature not present in any selected profiles? Finally, did clustering highlight overlapping helix classes that could coexist in the sampled structures? We labelled the clustering improved (Imprv) if at least one of these conditions was satisfied.

To illustrate, consider again the *D. mobilis* 5S rRNA sequence. By comparing the summary profile graph (Fig. 7a) with the clustering results, we see that clustering retains information about feature 154, which did not appear in any selected profile. One of the clusters (Summary 4 in Fig. 9) shows that the overlapping features 154 and 34 may coexist. Interestingly, all but one of the structures having feature 154 also contain feature 34.

Alternatively, consider the *V. cholerae* Qrr3 sequence. As discussed, one cluster (Summary 0 in Fig. 10) contained the two selected profiles [[1][13[20]][10][0]] and [[1][13[20]][0]]. Further inspection revealed that most (74 out of 93) of the extended profiles not containing feature 10 instead contained less frequent helix classes $5 = (46, 71, 8)$, $12 = (49, 72, 7)$, and/or $27 = (52, 63, 4)$, which coincide with the same general region of base pairs as feature 10. Thus, the information in the extended profiles showed that both of these selected profiles actually have similar base pairings, and it is more informative to consider them in one group of structures.

Overall, the new approach was judged to provide more structural information in 75% of the test cases, and to provide at least as much in 2 of the 3 remaining ones.

```
Sequence: ACGGUGCCCGACCCGGCCAUAGUGGC
CGGGCAACACCCGGUCUCGUUUCGAACCCGGAAGUU
AAGCCGGCCACGUCAGAACGGCCGUGAGGUCCGAGA
GGCCUCGCAGCCGUUCUGAGCUGGGAUCGGGCACC
CH-Index: 164354.62111801244


partition id, multiplicity,
structure helices
----------
Summary: 0, 346, 0 1 4 5 9 34 151 212
256 1249
0, 344, 0 1 4 5 9 34 151 212 256
0, 2, 0 1 4 5 9 34 151 212 256 1249
----------
Summary: 1, 201, 0 1 4 5 9 34 151 212
1, 201, 0 1 4 5 9 34 151 212
----------
Summary: 2, 173, 0 1 4 5 9 34 104 151
212 256
2, 173, 0 1 4 5 9 34 104 151 212 256
----------
Summary: 3, 104, 0 1 4 5 9 34 104 151 212
3, 104, 0 1 4 5 9 34 104 151 212
----------
Summary: 4, 62, 0 1 4 5 9 34 151 154
212 256 318 1249
4, 1, 0 1 4 5 9 34 151 154 212 256 1249
4, 1, 0 1 4 5 9 34 151 154 212 256 318
4, 60, 0 1 4 5 9 34 151 154 212 256
----------
Summary: 5, 29, 0 1 4 5 9 34 151 154 212
5, 29, 0 1 4 5 9 34 151 154 212
----------
Summary: 6, 28, 0 1 4 5 9 34 104 151
154 212 256 1249
6, 27, 0 1 4 5 9 34 104 151 154 212 256
6, 1, 0 1 4 5 9 34 104 151 154 212
256 1249
----------
Summary: 7, 12, 0 1 4 5 9 34 104 151 154 212
7, 12, 0 1 4 5 9 34 104 151 154 212
----------
Summary: 8, 10, 0 1 4 5 9 34 151 256 318 461
8, 8, 0 1 4 5 9 34 151 256
8, 1, 0 1 4 5 9 34 151 256 461
8, 1, 0 1 4 5 9 34 151 256 318
----------
```

```
Summary: 9, 8, 0 1 4 5 9 34 151 461
9, 6, 0 1 4 5 9 34 151
9, 2, 0 1 4 5 9 34 151 461
----------
Summary: 10, 6, 0 1 4 5 9 34 104 151 256 461
10, 2, 0 1 4 5 9 34 104 151 256 461
10, 4, 0 1 4 5 9 34 104 151 256
----------
Summary: 11, 5, 0 1 4 5 9 34 212 256
455 1249
11, 1, 0 1 4 5 9 34 212 256 455 1249
11, 2, 0 1 4 5 9 34 212 256 455
11, 2, 0 1 4 5 9 34 212 256
----------
Summary: 12, 4, 0 1 4 5 9 34 104 151 461
12, 1, 0 1 4 5 9 34 104 151
12, 3, 0 1 4 5 9 34 104 151 461
----------
Summary: 13, 4, 0 1 4 5 9 34 151 154 256 461
13, 1, 0 1 4 5 9 34 151 154 256 461
13, 3, 0 1 4 5 9 34 151 154 256
----------
Summary: 14, 1, 0 1 4 5 9 34 104 151 154
14, 1, 0 1 4 5 9 34 104 151 154
----------
Summary: 15, 1, 0 1 4 5 9 34 104 154
212 256 455
15, 1, 0 1 4 5 9 34 104 154 212 256 455
----------
Summary: 16, 1, 0 1 4 5 9 34 104 212
16, 1, 0 1 4 5 9 34 104 212
----------
Summary: 17, 1, 0 1 4 5 9 34 104 212 256 455
17, 1, 0 1 4 5 9 34 104 212 256 455
----------
Summary: 18, 1, 0 1 4 5 9 34 151 154 461
18, 1, 0 1 4 5 9 34 151 154 461
----------
Summary: 19, 1, 0 1 4 5 9 34 154 212 455
19, 1, 0 1 4 5 9 34 154 212 455
----------
Summary: 20, 1, 0 1 4 5 9 34 455
20, 1, 0 1 4 5 9 34 455
----------
Summary: 21, 1, 0 1 4 5 9 104 151 154
212 256 318
21, 1, 0 1 4 5 9 104 151 154 212 256 318
```

Fig. 9 Output file of the partition for the *D. mobilis* 5S rRNA extended profiles at threshold 96 corresponding to the clusters visualized in Fig. 7b on page 66

Of the five "easy" sequences, the graph clustering approach provided additional structural information for three (*V. cholerae* Qrr3, *B. clausii* TPP, and *D. mobilis* 5S rRNA), and recapitulated one (*C. botulinu* THF) but did neither for *V. cholerae* Qrr1. Of the four sequences classified as medium difficulty for profiling, three improved (*V. harveyi* Qrr1, *P. multocid* TPP, *A. tabira* 5S rRNA) while one (*S. uberis* THF) was just recapitulated. Finally, there were three sequences which were hard for profiling to analyze, making it more difficult to compare and contrast the outcomes. However, the clustering output for all three (*M. multacida* THF, *T. acidophilum* TPP, *E. coli* 5S rRNA) met the criteria for improvement.

```
Sequence: UGACCCUUAAUUAAGCCGAGGGUCAC       0, 1, 0 1 13 20 21 27 170
CUAGCCAACUGACGUUGUUAGUGAAUGAAAUUGUUC       0, 1, 0 1 13 20 21 27
ACAUUUGUUUUAUCAGCCAAUCACCCUUUUGUGAUU       0, 1, 0 1 12 13 20 34 49
GGCUUUUUU                                  0, 1, 0 1 12 13 20 34
CH-Index: 1615.9213625205339              0, 1, 0 1 10 13 20 595
                                          0, 1, 0 1 10 13 20 169
                                          0, 1, 0 1 10 13 20 55 158
partition id, multiplicity,               0, 1, 0 1 10 13 20 54
structure helices                         0, 1, 0 1 10 13 20 21 402
----------                                0, 1, 0 1 5 13 20 281
Summary: 0, 507, 0 1 5 10 12 13           0, 1, 0 1 5 13 20 90
20 21 27 34 35 36 48 49 54 55             0, 1, 0 1 5 13 20 27 90
90 158 169 170 171 202 267 269            ----------
273 281 402 413 595 752                   Summary: 1, 425, 0 1 2 13 20 21
0, 247, 0 1 10 13 20                       33 48 90 267 390 391
0, 52, 0 1 10 13 20 49                     1, 393, 0 1 2 13 20
0, 44, 0 1 10 13 20 55                     1, 2, 0 1 2 13 20 267
0, 32, 0 1 10 13 20 402                    1, 1, 0 1 2 13 20 33
0, 24, 0 1 13 20 27                        1, 14, 0 1 2 13 20 90
0, 23, 0 1 5 13 20                         1, 7, 0 1 2 13 20 48
0, 9, 0 1 10 13 20 49 55                   1, 4, 0 1 2 13 20 21
0, 7, 0 1 12 13 20                         1, 3, 0 1 2 13 20 390
0, 6, 0 1 13 20                            1, 1, 0 1 2 13 20 391
0, 5, 0 1 10 13 20 752                     ----------
0, 5, 0 1 10 13 20 21 49                   Summary: 2, 52, 0 1 2 20 90 109
0, 5, 0 1 5 13 20 27                       2, 3, 0 1 2 20 109
0, 4, 0 1 10 13 20 55 402                  2, 3, 0 1 2 20 90
0, 4, 0 1 10 13 20 21                      2, 46, 0 1 2 20
0, 3, 0 1 12 13 20 49                      ----------
0, 3, 0 1 10 13 20 158                     Summary: 3, 7, 0 1 6 10 196 402
0, 2, 0 1 13 20 158                        3, 1, 0 1 6 10
0, 2, 0 1 13 20 34 35 49                   3, 3, 0 1 6 10 402
0, 2, 0 1 13 20 27 36                      3, 3, 0 1 6 10 196
0, 2, 0 1 10 13 20 48                      ----------
0, 2, 0 1 10 13 20 21 55                   Summary: 4, 6, 0 1 2 6 66 70 90 595
0, 1, 0 1 13 20 202                        4, 1, 0 1 2 6
0, 1, 0 1 13 20 90                         4, 1, 0 1 2 70 90
0, 1, 0 1 13 20 54 269                     4, 1, 0 1 2 66 595
0, 1, 0 1 13 20 49 273 413                 4, 3, 0 1 2 66
0, 1, 0 1 13 20 49 171                     ----------
0, 1, 0 1 13 20 35 49 273                  Summary: 5, 2, 0 1 10 13 49
0, 1, 0 1 13 20 34 49 170                  5, 1, 0 1 10 13
0, 1, 0 1 13 20 34 35                      5, 1, 0 1 10 13 49
0, 1, 0 1 13 20 34                         ----------
0, 1, 0 1 13 20 27 267                     Summary: 6, 1, 0 1 10 20 305
0, 1, 0 1 13 20 27 90 169                  6, 1, 0 1 10 20 305
0, 1, 0 1 13 20 27 90
```

Fig. 10 Output file of the partition for the *V. cholerae* Qrr3 sRNA extended profiles at threshold 67 corresponding to the clusters visualized in Fig. 8b on page 67

5 Conclusions

Based on these preliminary results, we conclude that this proof-of-principle analysis supports efforts to extract additional structural information from the Boltzmann sample beyond the features and selected profiles identified by profiling. In moving forward, several factors are worth consideration.

One option we considered, but did not implement, was using an edit distance as the (dis)similarity weighting. This was motivated by the problem of multiple competing helices which produce similar structures, as for *V. harveyi* Qrr1 (a non-improving sequence). An edit distance similarity score might also simplify the analysis of hard sequences, like *E. coli* 5S rRNA. Under more careful consideration, however, the difficulty became how to implement it efficiently; if the edit distance score involves a matching of helices, then it resembles the *NP*-hard problem of *Assignment Problem with Disjunctive Constraints* [18]. It is possible that our problem can be formulated to avoid this complexity issue, but further exploration of this option was beyond the scope of this initial study.

We also note that it will be important to consider how the clusters are presented. The original method, by design, yields an interpretable signature—the profile— for each equivalence class in the partition. It is not immediately obvious what the corresponding label might be for the new partitions. While the centroid structure can be computed (as is done for the CH index), this is not necessarily the most informative way to represent a cluster.

Finally, there is the question of how to compare/contrast the different clusters. Profiling provides the summary profile graph for this purpose, which works well for the easy and even medium sequences. (A caveat worth mentioning, though, is that the transitive reduction used is the most computationally intensive profiling component.) In contrast, in our experience, the standard graph layout algorithms available are not well suited to visualizing the new community detection partitions. A visualization method which highlights the similarities and differences between the parts of the partition, yet is computationally tractable, especially as sequence length scales, would be quite useful.

Acknowledgments The work described herein was initiated during the Collaborative Workshop for Women in Mathematical Biology hosted by the Institute for Pure and Applied Mathematics at the University of California, Los Angeles in June 2019. Funding for the workshop was provided by IPAM, the Association for Women in Mathematics' NSF ADVANCE "Career Advancement for Women Through Research-Focused Networks" (NSF-HRD 1500481) and the Society for Industrial and Applied Mathematics. The authors thank the organizers of the IPAM-WBIO workshop (Rebecca Segal, Blerta Shtylla, and Suzanne Sindi) for facilitating this research.

The authors thank the anonymous reviewers whose comments significantly improved the paper.

This research of Huijing Du was supported in part by NSF-DMS 1853636; of Margherita Maria Ferrari by the NSF-Simons Southeast Center for Mathematics and Biology (SCMB) through NSF-DMS 1764406 and Simons Foundation SFARI 594594; of Christine Heitsch by NIH R01GM126554; of Forrest Hurley by GBMF4560 (to Sullivan) and NSF-DMS 1344199 (to Heitsch); of Christine Mennicke by an NSF GRFP; of Blair D. Sullivan by the Gordon & Betty Moore Foundation's Data-Driven Discovery Initiative under Grant GBMF4560; and of Bin Xu by the Robert and Sara Lumpkins Endowment for Postdoctoral Fellows in Applied and Computational Mathematics and Statistics at the University of Notre Dame.

Appendix

Two data files (Figs. 9 and 10) are included to supplement the examples presented and discussed in Sect. 4.5. The files begin by listing the RNA sequence analyzed and the CH index for this partition, which had the largest proportional increase. Recall that the (dis)similarity measure is the frequency-weighted symmetric difference and the community detection is performed by the GMC graph algorithm.

The clusters of the partition are listed by decreasing total frequency. Each one is first described by a "Summary" which lists a cluster label ('partition id') and its total frequency mass, followed by the list of helix classes appearing in the union of the the cluster elements. This Summary is followed by a list of all the extended profiles in the cluster, also sorted by decreasing frequency. Each extended profile is specified by the cluster label, its frequency in the sample ('multiplicity'), and the set of helix classes particular to this extended profile.

References

1. Caliński, T., Harabasz, J.: A dendrite method for cluster analysis. Commun. Stat. - Theory Methods **3**(1), 1–27 (1974)
2. Chan, C.Y., Lawrence, C.E., Ding, Y.: Structure clustering features on the Sfold Web server. Bioinformatics **21**(20), 3926–3928 (2005)
3. Clauset, A., Newman, M.E.J., Moore, C.: Finding community structure in very large networks. Phys. Rev. E **70**, 066111 (2004)
4. Cordasco, G., Gargano, L.: Community detection via semi-synchronous label propagation algorithms. Int. J. Soc. Netw. Min. **1**, 3–26 (2012)
5. Crick, F.: Codon—anticodon pairing: The wobble hypothesis. J. Mol. Biol. **19**(2), 548–555 (1966)
6. Ding, Y., Chan, C.Y., Lawrence, C.E.: RNA secondary structure prediction by centroids in a Boltzmann weighted ensemble. RNA **11**(8), 1157–1166 (2005)
7. Ding, Y., Chan, C.Y., Lawrence, C.E.: Clustering of RNA secondary structures with application to messenger RNAs. J. Mol. Biol. **359**(3), 554–571 (2006)
8. Ding, Y., Lawrence, C.E.: A statistical sampling algorithm for RNA secondary structure prediction. Nucleic Acids Res. **31**(24), 7280–7301 (2003)
9. Giegerich, R., Voß, B., Rehmsmeier, M.: Abstract shapes of RNA. Nucleic Acids Res. **32**(16), 4843–4851 (2004)
10. Girvan, M., Newman, M.E.J.: Community structure in social and biological networks. Proc. Natl. Acad. Sci. U.S.A. **99**, 7821–7826 (2002)
11. Hagberg, A.A., Schult, D.A., Swart, P.J.: Exploring network structure, dynamics, and function using NetworkX. In: G. Varoquaux, T. Vaught, J. Millman (eds.) Proceedings of the 7th Python in Science Conference, pp. 11 – 15. Pasadena, CA USA (2008)
12. Huang, J., Voß, B.: Analysing RNA-kinetics based on folding space abstraction. BMC Bioinformatics **15**, 60 (2014)
13. Kerpedjiev, P., Hammer, S., Hofacker, I.L.: Forna (force-directed RNA): simple and effective online RNA secondary structure diagrams. Bioinformatics **31**(20), 3377–3379 (2015)
14. Lenz, D.H., Mok, K.C., Lilley, B.N., Kulkarni, R.V., Wingreen, N.S., Bassler, B.L.: The small RNA chaperone Hfq and multiple small RNAs control quorum sensing in *Vibrio harveyi* and *Vibrio cholerae*. Cell **117**(1), 69–82 (2004)
15. McCaskill, J.S.: The equilibrium partition function and base pair binding probabilities for RNA secondary structure. Biopolymers: Original Research on Biomolecules **29**(6-7), 1105–1119 (1990)
16. Nussinov, R., Pieczenik, G., Griggs, J.R., Kleitman, D.J.: Algorithms for loop matchings. SIAM J. Appl. Math. **35**(1), 68–82 (1978)
17. Pérez-Reytor, D.e.a.: Role of non-coding regulatory RNA in the virulence of human pathogenic *Vibrios*. Front. Microbiol. **7** (2017)

18. Pferschy, U., Schauer, J.: The maximum flow problem with disjunctive constraints. J. Comb. Optim. **26**(1), 109–119 (2013)
19. Rogers, E., Heitsch, C.E.: Profiling small RNA reveals multimodal substructural signals in a Boltzmann ensemble. Nucleic Acids Res. **42**(22), e171–e171 (2014)
20. Schaeffer, S.E.: Graph clustering. Computer Science Review **1**, 27–64 (2007)
21. Shannon, C.E.: A mathematical theory of communication. Bell System Technical Journal **27**(3), 379–423 (1948)
22. Steffen, P., Voß, B., Rehmsmeier, M., Reeder, J., Giegerich, R.: RNAshapes: an integrated RNA analysis package based on abstract shapes. Bioinformatics **22**(4), 500–503 (2005)
23. Tu, K.C., Bassler, B.L.: Multiple small RNAs act additively to integrate sensory information and control quorum sensing in *Vibrio harveyi*. Genes Dev **21**, 221–233 (2007)
24. Turner, D.H., Mathews, D.H.: NNDB: the nearest neighbor parameter database for predicting stability of nucleic acid secondary structure. Nucleic Acids Res **38**(suppl 1), D280–D282 (2010)
25. Turner, D.H., Sugimoto, N., Freier, S.M.: RNA structure prediction. Annu. Rev. Biophys. Biophys. Chem. **17**(1), 167–192 (1988)
26. Waterman, M.S., Smith, T.F.: RNA secondary structure: A complete mathematical analysis. Math. Biosci. **42**(3-4), 257–266 (1978)
27. Watson, J.D., Crick, F.H., et al.: Molecular structure of nucleic acids. Nature **171**(4356), 737–738 (1953)
28. Wuchty, S., Fontana, W., Hofacker, I.L., Schuster, P.: Complete suboptimal folding of RNA and the stability of secondary structures. Biopolymers: Original Research on Biomolecules **49**(2), 145–165 (1999)
29. Zuker, M.: Computer prediction of RNA structure. In: Methods in enzymology, vol. 180, pp. 262–288. Elsevier (1989)
30. Zuker, M.: On finding all suboptimal foldings of an RNA molecule. Science **244**(4900), 48–52 (1989)
31. Zuker, M., Mathews, D.H., Turner, D.H.: Algorithms and thermodynamics for RNA secondary structure prediction: a practical guide. In: RNA Biochem Biotechnol, pp. 11–43. Springer (1999)
32. Zuker, M., Stiegler, P.: Optimal computer folding of large RNA sequences using thermodynamics and auxiliary information. Nucleic Acids Res. **9**(1), 133–148 (1981)

How Do Interventions Impact Malaria Dynamics Between Neighboring Countries? A Case Study with Botswana and Zimbabwe

Folashade Agusto, Amy Goldberg, Omayra Ortega, Joan Ponce, Sofya Zaytseva, Suzanne Sindi, and Sally Blower

Abstract Malaria is a vector-borne disease that is responsible for over 400,000 deaths per year. Although countries around the world have taken measures to decrease the incidence of malaria, many regions remain endemic. Indeed, progress towards elimination has stalled in multiple countries. While control efforts are largely focused at the national level, the movement of individuals between countries may complicate the efficacy of elimination efforts. Here, we consider the case of neighboring countries Botswana and Zimbabwe, connected by human mobility. Both have improved malaria interventions in recent years with differing success. We use a two-patch Ross-MacDonald model with Lagrangian human mobility to examine the coupled disease dynamics between these two countries. In particular,

F. Agusto
University of Kansas, Lawrence, KS, USA

A. Goldberg
Duke University, Durham, NC, USA
e-mail: amy.goldberg@duke.edu

O. Ortega
Sonoma State University, Rohnert Park, CA, USA
e-mail: ortegao@sonoma.edu

J. Ponce
Purdue University, West Lafayette, IN, USA
e-mail: ponce0@purdue.edu

S. Zaytseva
University of Georgia, Athens, GA, USA
e-mail: szaytseva@uga.edu

S. Sindi (✉)
Department of Applied Mathematics, University of California, Merced, CA, USA
e-mail: ssindi@ucmerced.edu

S. Blower (✉)
University of California Los Angeles, Los Angeles, CA, USA
e-mail: SBlower@mednet.ucla.edu

© The Association for Women in Mathematics and the Author(s) 2021
R. Segal et al. (eds.), *Using Mathematics to Understand Biological Complexity*,
Association for Women in Mathematics Series 22,
https://doi.org/10.1007/978-3-030-57129-0_5

83

we are interested in the impact that interventions for controlling malaria applied in one country can have on the incidence of malaria in the other country. We find that dynamics and interventions in Zimbabwe can dramatically influence pathways to elimination in Botswana, largely driven by Zimbabwe's population size and larger basic reproduction number.

Keywords Malaria · Multi-patch model · Epidemiology · Vector disease transmission · Basic reproduction number · Ordinary differential equations

1 Introduction

Concerted efforts over the past 20 years have dramatically decreased the incidence of malaria in many countries around the world. However, the response to interventions to reduce malaria has varied geographically, with neighboring countries' efforts often producing significantly different results. For example, in Botswana, from 2000 to 2012, annual malaria cases were reduced from over 70,000 to only about 300. While neighboring countries Zimbabwe, Namibia, and Zambia have also decreased their malaria rates, they remain high-infection regions, with substantial tourism and migration connecting these countries and Botswana [51, 53, 57, 72].

Recent empirical approaches have demonstrated the role that source-sink dynamics can play in maintaining epidemics in regions that would not sustain disease transmission in isolation [40, 56]. Here, we use a model-based approach to examine the role that human movement can play in infection dynamics in regions that are interconnected by human mobility, and that are close to eliminating the disease. For simplicity, we use Botswana and Zimbabwe as a case study to consider infection dynamics as both of these countries attempt to go from low infection rates to elimination while remaining connected by human movement.

While a variety of models have considered the infection rates in malaria endemic countries [2, 41, 67, 69], little is understood about the final steps before elimination. As more countries move closer to malaria elimination, it is important to understand the dynamics of infection when the number of cases is low. This period of endemicity is particularly important because of recent empirical observations that infection rates have increased in multiple countries that were previously on positive trajectories towards elimination [50, 57]. Many models of malaria do not include human movement, yet it has been an important factor preventing elimination in many countries. For example, malaria was re-introduced to Greece and Sri Lanka through migration [43, 58], and Botswana has seen an uptick in infection rates since 2017, including an increase in imported cases [57].

Studies that considered movement have often considered a two-patch model of human and vector dynamics in the context of malaria transmission. Cosner et al. [23] demonstrated that movement between humans is important for disease persistence. In their study, the authors built a two-patch model in which the disease would have died out in both patches in isolation, but is sustained by human movement. The model uses two different descriptions of movement. In the first type of movement, humans are residents of a given patch and remain in that patch most of the time, while occasionally visiting other patches often enough for pathogen transmission to occur in the visited patch. The infection rate for humans in a given patch depends on the number of infectious vectors in other patches and the fraction of the time that a visiting individual spends in those patches. However, the infection rate is not tied to any specific type of human movement between classes or patches. Acevedo et al. [1] studied the impact of human migration in a multi-patch model. They assumed that the rate of host movement was symmetric between any two patches, and equal amongst all patches, for simplicity. They showed that local transmission rates are highly heterogeneous, and the reproduction number, R_0, declines asymptotically as human mobility increases. Ruktanonchai et al. [62, 63] also studied the impact of human mobility on malaria. The authors conducted an extensive theoretical study of the system level R_0 under a multi-patch model, and considered how malaria could be eliminated [63]. They also characterized mobility with call-records from mobile phones to determine transmission foci [62]. Prosper et al. [59] showed that even regions with low malaria transmission connected by human movement to regions with higher malaria endemicity should engage in malaria control programs. However, these previous studies largely focused only on the asymptotic elimination of the disease by reducing the system basic reproduction number, and not on the dynamics of the disease from the time intervention begins.

We use a multi-patch model to identify processes that could hinder elimination prospects, focusing on migration from other endemic countries. Specifically, we hypothesize that migration from malaria-endemic neighbors, particularly Zimbabwe, is a barrier to elimination of malaria in Botswana. To test this hypothesis, we use multi-patch Ross-MacDonald models [23, 62, 63]. In a multi-patch environment, individual patches may be characterized as sinks or sources [33, 40, 56]. Sinks are characterized by having low transmission (single-patch $R_0 < 1$), not enough to sustain an epidemic. Sources are characterized by high transmission (single-patch $R_0 > 1$), enough to sustain an epidemic in isolation. We investigate the role that different intervention strategies and mobility patterns can play on the source-sink dynamics of our multi-patch system. In contrast to most previous studies, we consider both the R_0 and the number of infections in each patch. We study these quantities in Botswana under varying migration rates from neighboring Zimbabwe, and use elasticity analysis to identify the most effective intervention strategies.

Resources for interventions to reduce malaria are limited, and often directed at a single-country level. Therefore it is important to understand the relative utility of various interventions types and locations. Such intervention strategies may have different relative effectiveness under different regimes of population density or migration. Under our model, we test which interventions, in which patch, may be most effective in reducing malaria in Botswana. Considering source-sink dynamics of the system, we examine how interventions in one patch influence the infection rate in the other patch.

In Sect. 2, we first provide details on Botswana and Zimbabwe, our two-country case study. We then present the model of malaria dynamics we are using and detail the metrics and parameters we use for Botswana and Zimbabwe. In Sect. 3, we study the dynamics of a two-patch model under different scenarios between the two countries. In Sect. 4, we discuss our findings and generalizations of our approach.

2   Malaria Dynamics in Botswana and Zimbabwe

In this case study we consider malaria dynamics between the connected countries Botswana and Zimbabwe. The malaria burden in Botswana is low, but potentially increasing, as it is surrounded by highly malaria-endemic countries. The areas that report the highest malaria burden are located in northern Botswana, including the Okavango delta, Ngamiland and Chobe, and to some extent Boteti and Tutume [51, 53, 72]. We focus on the first three regions: the Okavango delta, Ngamiland and Chobe. Interestingly, they do not contain the majority of Botswana's population. Instead, most people reside along the Eastern side of the country due to better environmental conditions such as more frequent rains and fertile soil [30]. However, our focus areas are located on the borders with Zimbabwe, Zambia, and Namibia, and include the majority of the malaria cases as well as some of the busiest border posts (Fig. 2). As more than 93% of all arrivals into Botswana occur by road [72], the transmission of malaria through these ports of entry from areas of higher malaria incidence into Botswana requires further investigation.

Multiple countries border Botswana and may influence malaria dynamics. Here, we focus on Zimbabwe as a malaria-endemic neighbor to Botswana for two reasons. First, according to the official statistics for 2017 [14], Zimbabwe is home to the majority of people traveling into Botswana on an annual basis. Second, Zimbabwe continues to be a highly malaria endemic country, with overall larger malaria incidence (defined as number of cases of the disease, per person per year) as compared to Botswana (see Figs. 1 and 2).

Fig. 1   Malaria Incidence in Botswana and Zimbabwe. Yearly malaria incidence for Botswana and Zimbabwe based on World Health Organization report in 2018 [57]

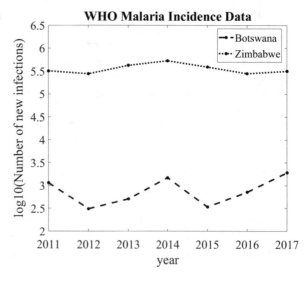

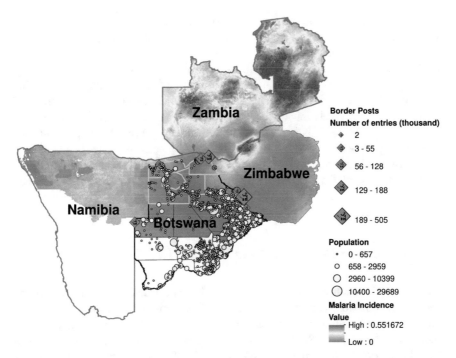

Fig. 2 Malaria Incidence in Botswana and Neighboring Countries. Map of Botswana with a subset of its neighbors, with spatial interpolation of malaria incidence (number of cases per person per year) [11]. For Botswana, we also plot the population distribution [55], and relevant border posts with annual number of entries [14]. We see that while Botswana has the lowest malaria incidence of all its neighbors, the risk of malaria transmission from the bordering countries is high given the number of border posts and the number of yearly arrivals from the neighboring countries

Human movement is often considered under two different frameworks: Eulerian (migration) and Lagrangian (visitation) movement. Eulerian movement involves migration of individuals between patches. Usually, these individuals do not reside in these patches but move freely between patches [63]. This movement approach assumes individuals become infected within the patches [23]. Most models involving mobility use this approach to depict movement from one patch to another [3, 5–7, 23, 26, 29, 34, 38, 46, 47, 67, 68, 75, 76]. Lagrangian movement on the other hand involves individuals residing in a specific patch and taking short commute or visits between patches, and spending part of their time away from their home patch. In the cause of these short visits are exposed to or can infect others with the pathogen [23, 63]. This type of movement has been used in a number of previous studies [23, 27, 37, 60, 61, 63].

Here, we focus on Lagrangian movement for a few reasons. Due to an economic and political down turn, Zimbabweans have been increasingly crossing into Botswana for work in cow herding, construction, real estate, retail, education, health, manufacturing, and for visits or holidays since the early 2000s [16, 44].

Some people cross to shop or trade [16]. Some of these cross-border migrant have up to 90 days of legal stay in Botswana, and those crossing to shop or to trade stay for even shorter periods [16].

However, legal immigration into Botswana has been on the decline according to the national census [15], with less than 0.2% of the total population being foreign workers, with valid worker permits [15]. Similarly, we expect undocumented migration to be relatively low compared to visitation because of recently introduced heightened border controls and increased punishment measures aimed to curb the number of people entering into Botswana illegally, particularly from Zimbabwe. Therefore, while permanent migration in and out of Botswana is present, we first focus on the simpler model with visitation-only (temporary) movement between patches.

2.1 Two-Patch Botswana-Zimbabwe Model

Mathematical models of malaria transmission have provided insight into the factors driving transmission, and the effectiveness of possible interventions, which have formed the basis of predictions under scenarios of climatic, cultural or socio-economic change [42, 48, 78]. We follow one of the most prominent models of malaria transmission, the deterministic coupled differential equations of the Ross-MacDonald model. These equations consider the infection rates of humans and mosquitoes over time as a function of human recovery rate, mosquito ecology, human and mosquito population sizes, and human-mosquito interactions [48, 66].

To study malaria dynamics in Botswana and Zimbabwe, we use the two-patch model of [23, 63]. Within each patch, the dynamics are governed by the (one-patch) Ross-MacDonald equations. Individuals live in one patch/country, but may spend some proportion of their time in the other patch/country (Fig. 3). To spatially couple the two patches, we follow the Lagrangian approach and assume that the movement

Fig. 3 Conceptual Two-Patch Malaria Model. Patch 1 (Botswana) and Patch 2 (Zimbabwe) contain both infected humans (X_1 and X_2) and infected mosquitoes (Y_1 and Y_2). Interactions that could result in infection are identified with dotted lines. Thick dotted lines denote within patch routes of infection, while the thin dotted lines denotes infection acquired by human mobility

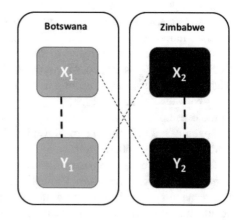

dynamics between patches is predominantly characterized by visitation, as opposed to permanent migration. In this way, we incorporate the fraction of time that the infected population of both mosquitoes and humans in patch 1 spends in patch 2 and vice versa [23]. We let the infected human populations from patch i, be X_i, and infected mosquitoes from patch i, be Y_i. Additionally, we make the assumption that our total human population (H_i) is fixed at steady state – to simplify our calculations – and allow X_i, the number of infected humans, and Y_i, the number of infected mosquitoes, to vary. Coupling the dynamics in both patches, the two-patch malaria model with Lagrangian movement is given as:

$$\frac{dX_1}{dt} = \left(p_{11}a_1b_1e^{-\mu_1\tau_1}Y_1 + p_{12}a_2b_2e^{-\mu_2\tau_2}Y_2\right)\frac{(H_1 - X_1)}{H_1} - r_1X_1 \quad (1)$$

$$\frac{dX_2}{dt} = \left(p_{21}a_1b_1e^{-\mu_1\tau_1}Y_1 + p_{22}a_2b_2e^{-\mu_2\tau_2}Y_2\right)\frac{(H_2 - X_2)}{H_2} - r_2X_2 \quad (2)$$

$$\frac{dY_1}{dt} = \left(q_{11}a_1c_1\frac{X_1}{H_1} + q_{12}a_2c_2\frac{X_2}{H_2}\right)(V_1 - Y_1) - \mu_1Y_1 \quad (3)$$

$$\frac{dY_2}{dt} = \left(q_{21}a_1c_1\frac{X_1}{H_1} + q_{22}a_2c_2\frac{X_2}{H_2}\right)(V_2 - Y_2) - \mu_2Y_2. \quad (4)$$

Table 1 Parameters in the model. Definitions correspond to the Ross-MacDonald model with Lagrangian dynamics (Equations (1), (2), (3), and (4)). We designate patch $i = 1$ as Botswana and patch $i = 2$ as Zimbabwe. Parameter values are taken from the literature, except a, which is fit to R_0 values. In certain analyses, parameters a and μ are varied to study intervention strategies. Further described in Sect. 2.2

Parameter	Value, Botswana	Value, Zimbabwe	Definition
R_0	1.01	1.5	Median reported value by Malaria Atlas Project
H_i	175,631	12,973,808	Total human population in patch i at equilibrium values
V_i	$10H_1$	$10H_2$	Total mosquito population
a_i	0.082	0.241	Human biting rate of mosquitoes in patch i
b_i	0.5	0.5	Transmission efficiency from infected mosquitoes to humans
c_i	0.1	0.1	Transmission efficiency from infected humans to mosquitoes
μ_i	1/30	1/10	Mosquito mortality rate patch i
τ_i	10	10	Incubation period; time a mosquito becomes infected until it becomes infectious
r_i	1/14	1/14	Recovery rate of humans in patch i
p_{ij}	$\in [0, 1]$	$\in [0, 1]$	Fraction of time a human resident in patch i spends visiting patch j
q_{ij}	$\in [0, 1]$	$\in [0, 1]$	Fraction of time a mosquito resident in patch i spends visiting patch j

(The parameter and variable definitions are given in Table 1.) The model incorporates human movement through the visitation parameters p_{ij}, defined as the proportion of time an individual from population i spends in population j. For simplicity, we assume that mosquitoes do not move. That is, we fix $q_{11} = q_{22} = 1$ and $q_{12} = q_{21} = 0$.

In our analysis of coupled malaria dynamics in Botswana and Zimbabwe we consider two metrics: (1) R_0, the Basic Reproduction Number (both at the system level and single-patch level), (2) the number of new infections per year in each patch. We next describe these quantities in terms of our model.

2.1.1 The Reproduction Number R_0

The basic reproduction number, R_0, represents the average number of secondary infections from an infected individual. Generally, when $R_0 > 1$, then infection will spread, and when $R_0 < 1$, infection will eventually decrease to zero. As such, R_0 is a metric that reflects the long-term asymptotic tendency of the infection dynamics. The approach to compute R_0, under the two-patch system is given in [23, 63]. The expression for the reproduction number, R_0, under single patch Ross-MacDonald model is given by [23, 48, 63]

$$R_0 = \left(\frac{ab}{\mu}\right)\left(\frac{\frac{V}{H}ace^{-\mu\tau}}{r}\right). \tag{5}$$

We see that R_0 is the product of the expected number of humans infected by a single infectious mosquito over its lifetime as well as the number of infected mosquitoes that arises from a single infectious human over the infection period.

Next, we consider the system level reproduction number R_0. While our subsequent model analysis focuses exclusively on the two-patch model, for generality we describe system level R_0 calculation in terms of the n-patch model. Consider the n-dimensional analogue of Equations (1), (2), (3), and (4). (In this case we have $n \times n$ matrices P and Q depicting the human and mosquito mobility respectively.) First, following [63], we rescale the model in each patch

$$\frac{dx_i}{dt} = \left(\sum_{j=1}^{n} \rho_i^{-1} p_{ij} m_{ij} a_j b_j e^{-\mu_j \tau_j} y_j\right)(1 - x_i) - r_i x_i \tag{6}$$

$$\frac{dy_i}{dt} = \left(\sum_{j=1}^{n} q_{ij} a_j c_j\right)(1 - y_i) - \mu_i y_i, \tag{7}$$

where $x_i = X_i/H_i$, $y_i = Y_i/V_i$, $m_{ij} = V_j/H_i$ is the ratio of the total number of mosquitoes over the total number of humans and $\rho_i = H_i/H_1$ is the ratio of the total human population in patch i and the total human population in the first patch.

Writing the system represented by Equations (6) and (7) in terms of the individual patch reproduction numbers R_0^j we have

$$\frac{dx_i}{dt} = \sum_{j=1}^{n} (\rho_i^{-1} p_{ij} \rho_j R_0^j \alpha_j^{-1} r_j y_j)(1 - x_i) - r_i x_i \tag{8}$$

$$\frac{dy_i}{dt} = \sum_{j=1}^{n} (q_{ij} \alpha_j \mu_j)(1 - y_i) - \mu_i y_i. \tag{9}$$

The global reproduction number for the n-patch model when both humans and mosquitoes move is given by the spectral radius of the matrix, $R(S)$, where S is given by

$$S = P \text{diag}(R_0) D^{-1} Q D,$$

where $D = \text{diag}((a_i c_i)/(r_i \rho_i))$ and $\text{diag}(R_0)$ is the matrix with the single patch R_0^j's on the diagonal.

The two-patch reproduction number (see below Equation (10)) is a special case of when only humans are moving. In this case, $Q = I$ and S simplifies to S_h:

$$S_h = P \text{diag}(R_0).$$

Recall that P is the matrix associated with human movement. Furthermore, the spectral radius $R(S_h)$ is bounded by the minimum and maximum single patch R_0's. That is,

$$(R_0)_{\min} \le R(S_h) \le (R_0)_{\max},$$

where $(R_0)_{\min} = \min_i(R_0^i)$, and $(R_0)_{\max} = \max_i(R_0^i)$. Note, we will obtain the same expression for the reproduction number if we use the standard approach of the next-generation matrix method [63, 74].

Using the theory described above, the system-level R_0 for a two-patch model can be written as

$$R_0 = \frac{p_{11} R_0^1}{2} + \frac{p_{22} R_0^2}{2} + \frac{\sqrt{(p_{11} R_0^1 + p_{22} R_0^2)^2 - 4(p_{11} + p_{22} - 1) R_0^1 R_0^2}}{2}. \tag{10}$$

The first two terms in Equation (10) are the weighted average of the individual reproduction numbers in each patch and the second term is the average number of secondary infections imported into each patch. The term under the square root is the average number of secondary infections imported into each patch.

Within the context of a multi-patch environment, individual patches are characterized as sinks (single-patch $R_0 < 1$) or sources (single-patch $R_0 > 1$) [56]. Based on Equation (10), if both patches in our two-patch model are sinks, the system R_0 will be less than 1 and malaria will asymptotically die out. If both patches are sources, then malaria will proliferate. In the following sections we consider the interesting case in which one patch is a sink and the other is a source.

2.1.2 The Number of New Cases

The second metric we use when evaluating malaria dynamics is the number of new infections in each patch i. That is, the total number of malaria infections for individuals in patch i regardless of where they were infected. We choose this particular metric as it allows for the comparison of the model output to data on the number of new malaria cases, commonly reported by such agencies as the World Health Organization (WHO) [57]. The first term in the X_i Equations (1) and (2) represents the rate per unit time of new infections of individuals from patch i. Since our unit of time is days, the total number of infections within a year starting at t_0 is given by

Yearly New Cases Patch 1

$$:= \int_{t_0}^{t_0+365} \left(p_{11} a_1 b_1 e^{-\mu_1 \tau_1} Y_1(t) + p_{12} a_2 b_2 e^{-\mu_2 \tau_2} Y_2(t) \right) \frac{(H_1 - X_1(t))}{H_1} dt$$

(11)

Yearly New Cases Patch 2

$$:= \int_{t_0}^{t_0+365} \left(p_{21} a_1 b_1 e^{-\mu_1 \tau_1} Y_1(t) + p_{22} a_2 b_2 e^{-\mu_2 \tau_2} Y_2(t) \right) \frac{(H_2 - X_2(t))}{H_2} dt.$$

(12)

Notice that the terms in the previous equations could be further distinguished between new cases that were acquired in the home patch (p_{11} and p_{22} terms) and those that were acquired in the other patch (p_{12} and p_{21} terms). Because humans do not die from malaria in our model formulation, it is possible for the same individual to be counted multiple times in the number of new cases because they could be infected more than once during a given year. In our analysis below, we will study Equations (11) and (12) both at the steady-state values for X_i and Y_i and in response to different intervention strategies.

2.2 Choosing Parameters for Each Country

The final step before our analysis is to select parameters. Our two patch model for malaria dynamics has many parameters (see Table 1) that in principle could differ between patches. However, because the reported data for each country was limited, the parameters could not be determined uniquely for each patch. As such, we selected parameters according to the following process.

First, we determined the human and vector populations. For the human population in each country, we used reported values for each as shown in Table 1. Because there were wildly varying ranges for the ratio of mosquitoes to humans, and the number of mosquitoes may vary by a factor of 10 between the wet and dry seasons, for simplicity we assumed a fixed ratio of 10 female mosquitoes per human [9, 52].

Next, there were a number of kinetic parameters we assumed were the same between both patches. The rate of recovery of humans from malaria, r, varied in the literature and typically corresponded to recovery without treatment [10, 22]. Because both Botswana and Zimbabwe are countries that have undertaken efforts to control malaria, we assumed infected individuals would have access to treatment and estimated that the typical infected period of a human would be 14 days ($r = 1/14$) for both countries. Reported values for the transmission efficiency of malaria between mosquitoes to humans, b, and humans to mosquitoes, c, also varied [13, 20, 21, 36]. We selected the representative values of $b = 0.5$ and $c = 0.1$ and assumed these did not vary between patches. For the value of τ, the incubation time between a mosquito acquiring malaria and becoming infectious, we chose 10 which is consistent with the reported value in [54].

Finally, the remaining two parameters a and μ were chosen to be different in the two patches based on the reported use of interventions in Botswana and Zimbabwe. The use of insecticidal treated bednets (ITN) is one of the more common intervention strategies. Interestingly, while the ITN coverage for Zimbabwe has increased since 2011, the actual usage has decreased by 11% points in recent years [70]. In comparison, this does not seem to be an issue for Botswana, where the usage of nets has increased since 2011 [17] due to aggressive campaigns undertaken by various agencies [73]. The parameter in our model which would reflect this type of intervention is the feeding rate, a. Further, the overall coverage of indoor residual spraying (IRS) has remained high (about 90%) for Zimbabwe [64]. At the same time, IRS has been a problem area for Botswana since 2011, remaining at around 70% as reported by the WHO and [30, 65], despite the 90% target. The parameter in our model which reflects this type of intervention is the mosquito death rate, μ. Therefore, when considering intervention in both of these countries, we focus on the present day scenario where Botswana has a relatively smaller μ_1 value (corresponding to smaller mosquito death rate due to insufficient spraying (IRS) coverage), while Zimbabwe has a relatively larger a_2 value (corresponding to higher feeding rate due to insufficient bed net (ITN) coverage). We assumed $\mu_1 = 1/30$ for Botswana and $\mu_2 = 1/10$ for Zimbabwe. The value of a was then fit so that each country had the same R_0 as the median reported value for each country by the Malaria Atlas Project ($R_0 = 1.01$ for Botswana and $R_0 = 1.5$ for Zimbabwe) [8]. (See Table 1 for a full list of parameters used in our work.) For our simulations, we investigated both 10% and 20% changes in intervention strategies (parameters a and μ) as possible scenarios that could be undertaken by the governments of Botswana and Zimbabwe. Based on recent reports on malarial intervention strategies in these respective countries, as well as cost benefit analysis of the scale-up of intervention

[77], we feel that both the 10% and 20% changes (in either the positive or negative direction) are adequate and realistic intervention scenarios that could either happen by worsening of conditions or in the case that either government steps up their intervention strategies.

3 Results

With our model and parameters for each country established, we next analyze malaria dynamics in Botswana and Zimbabwe under several scenarios. First, the impact of mobility alone on system level behavior is considered. Second, we consider the impact of intervention strategies in one country, while the other remains the same. Finally, we consider the impact of changes in both countries at the same time. We consider the synergistic impact of improved interventions in both countries as well as how a worsening of malaria conditions in one country can impact the ability of interventions in the other country to eliminate malaria.

3.1 Impact of Mobility Alone on Botswana and Zimbabwe

We first focus on how mobility alone impacts malaria in our two-patch model under our two metrics. First, we consider the system level R_0. Because both countries have an R_0 value larger than 1 (Table 1), they are currently both sources. In this case, the mobility parameters in Equation (10) cannot drive the system R_0 below 1. However, mobility can cause the system level R_0 to be lower than the maximum single-patch R_0 (Zimbabwe) by increasing the amount of time individuals spend in patch with the lower R_0 (Botswana) (see Fig. 4 and Equation (10)). We note that, as expected, the system level R_0 depends more on p_{21} than p_{12} because of the larger single-patch R_0 of Zimbabwe.

The reproduction number is a consequence of the system parameters, so we next study its sensitivity to our choice of parameters. We use an elasticity analysis to gain insight into which parameters have the most impact on the basic reproduction number. The elasticity of the reproduction number R_0 to a general parameter p is simply the proportional change in R_0 resulting from a proportional change in p [19, 24, 59]:

$$\varepsilon_p = \frac{\delta R_0}{\delta p} \frac{p}{R_0}. \tag{13}$$

If the elasticity of R_0 with respect to a parameter p is ε_p, then a 1% change in p will result in an ε_p% change in R_0. That is, the elasticity gives the amount of change in R_0 in response to changes in p, making comparisons between parameters

Fig. 4 System Level R_0 Under Varying Mobility. When Botswana and Zimbabwe are both sources with reproduction numbers $R_0 = 1.01$ and $R_0 = 1.5$, respectively, the system level R_0 cannot be driven below 1 with mobility alone

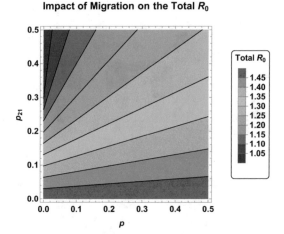

of different scales possible. Moreover, an elasticity analysis provides insight into prioritizing parameters for targeting by control strategies.

The elasticity analysis of the R_0 for each country individually identifies μ and a as the parameters with the largest impact on R_0 in both Botswana and Zimbabwe (Fig. 5a, b). Indeed, these two parameters are related to two most commonly implemented malaria interventions: indoor residual spraying and insecticide-treated bed nets. (We note that for the single patch R_0, the elasticity of parameters are similar between Botswana and Zimbabwe. This is to be expected because they share many parameters; however, differences appear in the elasticity for μ_i and τ_i.)

Next, we conduct an elasticity analysis of the system level R_0 for the two connected patches under two different mobility strategies. Figure 5c, d plot the elasticity of the system level R_0. We see that μ and a are still the parameters that most affect R_0. However, when we allow visitation, the extent of mobility, measured as p_{ij}, impacts the degree to which the system level R_0 is sensitive to the parameters. Therefore, we focus on a and μ for each country in conjunction with different mobility scenarios for the following sections. When the values for p_{12} and p_{21} are significantly different, the elasticity values change only in value but keep the same sign. For example, if p_{12} is significantly smaller than p_{21}, then the elasticity of the parameters with sub index 2 has larger elasticity values in magnitude preserving the same sign and vice versa. As above, the parameters in Zimbabwe (patch 2) all affect R_0 more than the analogous parameters in Botswana (patch 1).

Although mobility changes cannot eliminate malaria, we observe that they may substantially impact the number of cases of malaria in each patch at steady-state. In Fig. 6 we compare the number of cases of malaria under both high and low mobility between countries. (We fix the largest rate of mobility to be 0.5 since it is reasonable to assume that a resident would spend at least 50% of their time in their home patch.) We note that the number of cases overall is significantly lower when residents of Zimbabwe spend a large amount of time in Botswana. This makes

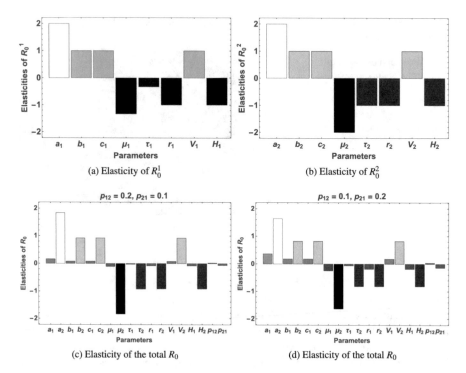

Fig. 5 Elasticity Analysis of the Basic Reproduction Number. (**a**) The elasticity of the R_0 in Botswana without visitation, $p_{12} = p_{21} = 0$. (**b**) The elasticity of the R_0 in Zimbabwe without visitation, $p_{12} = p_{21} = 0$. (**c**) The elasticity of the system level R_0 for the case of low mobility from Botswana to Zimbabwe and high mobility from Zimbabwe to Botswana, $p_{12} = 0.1$ and $p_{21} = 0.2$. (**d**) The elasticity of the system level R_0 for the case of high mobility from Botswana to Zimbabwe and low mobility from Zimbabwe to Botswana, $p_{12} = 0.2$ and $p_{21} = 0.1$. Darker colors indicate negative values, and lighter colors indicate positive values

sense as it exposes them to a more favorable R_0. For both patches and both high and low visitation rates to Botswana from Zimbabwe (p_{21}), we note that the more time a resident from Botswana spends in Zimbabwe (higher p_{12}) the higher the total number of cases is. As expected, the ratio of cases acquired locally compared to the total cases changes with p_{21}. For low p_{12} there appears to nearly always be a greater proportion of imported cases to Botswana while with high p_{12} it is possible for the local cases to exceed the imported cases for low p_{21}. In summary, these results show that while elimination is not possible, the more time any resident spends in the patch with the lower local reproduction number (in this case Botswana) the lower the total number of cases at steady-state.

Since changes in mobility alone are not sufficient to drive the system level R_0 below 1, we want to further investigate how changes in both intervention and mobility can significantly impact the overall disease dynamics. As both countries are still struggling to meet their malaria intervention goals, it is of interest how future

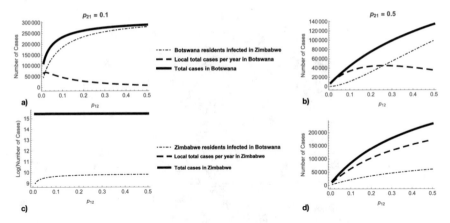

Fig. 6 Impact of Mobility on Local and Imported Malaria Cases. We fix the visitation rate of Zimbabwe residents visiting Botswana, $p_{21} = 0.1$ for (**a**) and (**c**) and $p_{21} = 0.5$ for (**b**) and (**d**), and vary the visitation rates of Botswana residents to Zimbabwe, p_{12}. We plot the steady-state number of infections per year in Botswana ((**a**) and (**b**)) and Zimbabwe ((**c**) and (**d**)). For both countries, there are substantially fewer cases under high rates of visitation from people in Zimbabwe to Botswana

changes in intervention strategies along with mobility patterns could influence malaria incidence in the region. Therefore, in the next sections, we consider four different scenarios: (1) Botswana improves its intervention strategy, Zimbabwe remains the same (2) Zimbabwe improves its intervention strategy, Botswana remains the same (3) Both countries improve their intervention strategies (4) Botswana improves its intervention strategy, Zimbabwe decreases the quality of its intervention.

3.2 Impact of a Successful Intervention Strategy in Botswana

Here, we investigate the impact of increased indoor residual spraying (IRS) in Botswana. As mentioned earlier, one of the challenges for Botswana remains to be a low uptake of vector control strategies. The implementation of indoor residual spraying has particularly been problematic, with IRS coverage consistently falling short of the 90% goal. Therefore, we argue that a realistic scenario for the future is the increase in spraying coverage in Botswana. In our model, this is controlled by the mosquito death rate μ_1. Therefore, an important question we ask is *if Botswana increases its IRS coverage and Zimbabwe does nothing, how much does mobility play a role in bringing down the system level R_0?*

Under this scenario, Botswana (a weak source), can easily be driven to be a sink with successful intervention ($R_0 < 1$) while Zimbabwe remains a source ($R_0 > 1$). From Fig. 7, we see that the total system R_0 can be brought down below 1 with

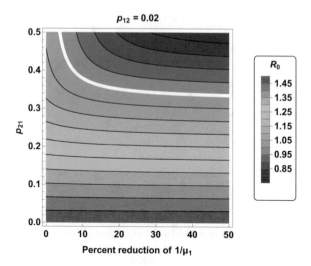

Fig. 7 System Level R_0 Under Intervention in Botswana. We fix $p_{12} = 0.02$, and allow p_{21} and $1/\mu_1$ to vary. Changes in μ_1 reflect changes in the insecticide spraying intervention in Botswana. The system total R_0 (Equation (10)) can be reduced below 1 by decreasing $1/\mu_1$. $R_0 = 1$ is shown in white

a combination of a modest increase in intervention in Botswana (starting with at least a 5% improvement) and a significant increase in mobility from Zimbabwe to Botswana (starting at 50% of the time a resident of Zimbabwe spends in Botswana).

Further, as Fig. 7 shows, the total system R_0 can also be reduced below 1, depicted by a white strip on the contour plot, by implementing more intervention and a less dramatic increase in mobility (p_{21}). However, for moderate to high improvements (at least 15%), the behavior is dominated by the mobility between Zimbabwe to Botswana (p_{21}).

Even if the R_0 is brought below 1, it may take a long time for the disease to die out. Therefore, we investigate the number of new infections over time in response to an intervention in Botswana (20% increase in μ_1), assuming implementation in 2019 under two different mobility scenarios. Under this intervention, Botswana is now a sink with a single-path $R_0 = 0.789$. As Fig. 8 shows, if mobility is high enough from Zimbabwe to Botswana, then malaria cases decrease. The level of mobility depicted, $p_{21} = 0.36$, was chosen to be just above the level that would drive the system R_0 below 1. However, we note that even 10 year later the number of cases in both countries is still far above 0.

Together, Figs. 7 and 8 demonstrate that if a successful intervention strategy can change a source country into a sink, the level of mobility between sink and source becomes important. In particular, it can result in the overall malaria elimination. However, elimination may still be years away.

Fig. 8 Impact of Mobility and Successful Intervention in Botswana on New Cases. The total number of new malaria cases (Equations (11) and (12)) under the scenario that Botswana increases μ_1 by 20% under two mobility strategies: high (dotted curves, Total $R_0 = 0.995$) and moderate (solid curves, Total $R_0 = 1.35$)

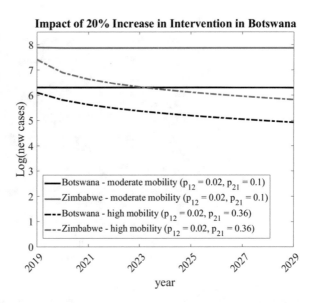

Impact of 20% Increase in Intervention in Botswana

— Botswana - moderate mobility ($p_{12} = 0.02$, $p_{21} = 0.1$)
— Zimbabwe - moderate mobility ($p_{12} = 0.02$, $p_{21} = 0.1$)
--- Botswana - high mobility ($p_{12} = 0.02$, $p_{21} = 0.36$)
--- Zimbabwe - high mobility ($p_{12} = 0.02$, $p_{21} = 0.36$)

3.3 Impact of a Successful Intervention Strategy in Zimbabwe

We next investigate increased bednet usage in Zimbabwe. As Zimbabwe continues to struggle with implementation of this intervention strategy, it is of interest how a more successful implementation of bednet usage can impact overall malaria dynamics in the entire region in the context of mobility. Therefore, an important question we ask is *if Zimbabwe increases its insecticide-treated bednet coverage (ITN) and Botswana does nothing, how much of a role does mobility play in bringing down the system level R_0?* In this case, we change the mosquito feeding rate a_2, which reflects changes in ITN. In this scenario, Zimbabwe can be driven to be a sink with successful intervention, while Botswana remains a weak source. We find that with significant improvement in bednet usage (at least a 20% improvement) and increased visitation from the patch with the higher R_0 (Botswana) to the patch with the lower R_0 (Zimbabwe), the system level R_0 can be decreased and brought down below 1 (see Fig. 9).

As before, we examine the dynamics of the new cases after this theoretical intervention begins in Zimbabwe (Fig. 10). Under this intervention, Zimbabwe is now a sink ($R_0 = 0.957$). As in Fig. 8, we observe that under the same reasonable mobility patterns ($p_{12} = 0.02$, $p_{21} = 0.1$) a successful intervention in Zimbabwe brings down the overall number of cases. However, when comparing the two intervention strategies (Figs. 8 and 10), we find that the intervention in Zimbabwe is more effective at reducing the total number of cases. The greater efficacy of the intervention in Zimbabwe makes sense as Zimbabwe has a larger population size. Further, as Fig. 10 shows, if only a 20% improvement in intervention is implemented (the necessary minimum for elimination) then mobility between Zimbabwe and

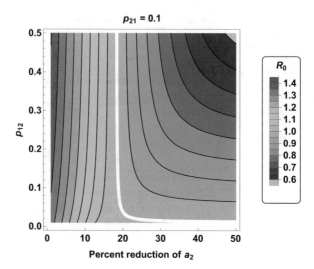

Fig. 9 System Level R_0 Under Intervention in Zimbabwe. Impact on the system total R_0 (Equation (10)) by varying visitation rate from Botswana to Zimbabwe and decreasing the human biting rate of the mosquitoes in Zimbabwe (a_2), corresponding to improvement in the insecticide-treated bednet coverage. The system $R_0 = 1$ is shown in white

Fig. 10 Impact of Mobility and Successful Intervention in Zimbabwe. The total number of new malaria cases (Equations (11) and (12)) under the scenario that Zimbabwe decreases a_2 by 20% under two mobility strategies: moderate (solid curves, system $R_0 = 1.005$) and low (dotted curves, system $R_0 = 0.998$)

Botswana actually has to be quite low to achieve elimination. The level of mobility depicted, $p_{21} = 0.02$, was chosen to be just below the level necessary to drive the total system R_0 below 1.

3.4 Synergistic Impact of Improved Interventions in Both Countries

Next, we consider the impact of an increase in intervention in both countries and ask the question *if both countries increase their intervention, how much does mobility play a role in bringing down the system level R_0?* In this case, we consider simultaneously changing μ_1 in Botswana and a_2 in Zimbabwe.

We find that improved intervention in both countries is a more viable option for elimination of the disease as it requires not only a less dramatic improvement in intervention on the part of both countries, but also a less dramatic change in mobility to obtain disease elimination. Figure 11 demonstrates that even under a moderate mobility scenario, $p_{21} = 0.1$, the disease may be eliminated with less effort on the parts of both countries.

We then investigate the number of new infections over time in response to a 10% improvement in intervention in both countries implemented in 2019 under two different mobility scenarios. Under this intervention, Botswana is a sink ($R_0 = 0.89$) and Zimbabwe remains a weaker source ($R_0 = 1.211$). As Fig. 12 shows, if mobility is high enough between Zimbabwe and Botswana, malaria can be eliminated. The level of mobility depicted, $p_{21} = .21$, was chosen just above the level necessary to drive the total system R_0 below 1. Comparing this with previous scenarios, we find that when both countries implement successful intervention strategies, it is possible to obtain disease elimination with an overall smaller improvement in intervention and a smaller change in mobility on the part of both countries. Together, Figs. 11 and 12 suggest that if both countries are able

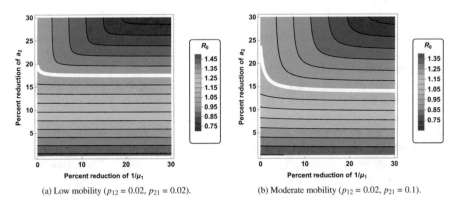

(a) Low mobility ($p_{12} = 0.02$, $p_{21} = 0.02$).   (b) Moderate mobility ($p_{12} = 0.02$, $p_{21} = 0.1$).

Fig. 11 System Level R_0 Under Intervention in Both Countries. (a) We show percent reduction of $\frac{1}{\mu_1}$ in Botswana on the x-axis, corresponding to decreasing the lifespan of a mosquito with successful usage of IRS (spraying) and percent reduction of a_2 in Zimbabwe (on the y-axis), corresponding to the decreasing of mosquito feeding rate through ITN intervention, under the scenario of low mobility. **(b)** This is the same as part **(a)**, but for the scenario of moderate mobility. The system $R_0 = 1$ is shown in white

Fig. 12 Synergistic Impact of Improved Interventions in Both Countries. The total number of new malaria cases (Equations (11) and (12)) under the scenario that Botswana increases μ_1 by 10% and Zimbabwe decreases a_2 by 10% under two mobility strategies: moderate (solid curves, system $R_0 = 1.1$) and high (dotted curves, system $R_0 = 0.994$)

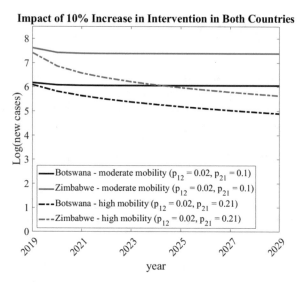

Impact of 10% Increase in Intervention in Both Countries

Botswana - moderate mobility ($p_{12} = 0.02$, $p_{21} = 0.1$)
Zimbabwe - moderate mobility ($p_{12} = 0.02$, $p_{21} = 0.1$)
Botswana - high mobility ($p_{12} = 0.02$, $p_{21} = 0.21$)
Zimbabwe - high mobility ($p_{12} = 0.02$, $p_{21} = 0.21$)

to make modest improvements, asymptotic elimination is more easily attained and requires a less dramatic change in mobility patterns.

3.5 Impact of a Worsening of Malaria Conditions in Zimbabwe on the Ability of Interventions in Botswana to Eliminate Malaria

Finally, we ask the question *if Botswana improves its intervention, while conditions in Zimbabwe become worse, how much does mobility play a role in bringing down the system level R_0?* Here, we again consider simultaneous changes in μ_1 for Botswana and changes a_2 for Zimbabwe which can drive Botswana to become a sink while Zimbabwe remains a strong source.

As this is a more extreme case of the first scenario discussed previously, we expect that the system level R_0 can be driven below 1 only if mobility from the source into the sink increases dramatically. Moreover, the increase in mobility has to be more significant than in the scenario where the conditions in Zimbabwe do not worsen. From Fig. 13, we see this is indeed the case. We again consider the dynamics of the new cases after this theoretical intervention. In this scenario, Botswana is driven to be a sink by a 20% increase in intervention ($R_0 = 0.789$), while Zimbabwe remains a strong source ($R_0 = 2.154$) with a 20% decrease in intervention. As Fig. 13 shows, if mobility is high enough between Zimbabwe and Botswana, malaria elimination can be achieved. The level of mobility necessary to result in elimination is $p_{21} = 0.58$, which is significantly higher than all other cases considered, and is an unrealistic scenario in which people spend more time away from their home

Fig. 13 Impact of a Worsening of Malaria Conditions in Zimbabwe and Interventions in Botswana on New Cases. The total number of new malaria cases (Equations (11) and (12)) under the scenario that Botswana increases μ_1 by 20% and Zimbabwe increases a_2 by 20% under two mobility strategies: moderate (solid curves, system $R_0 = 1.941$) and high (dotted curves, system $R_0 = 0.994$)

Impact of 20% Intervention Increase in Botswana and Decrease in Zimbabwe

- Botswana - moderate mobility ($p_{12} = 0.02$, $p_{21} = 0.1$)
- Zimbabwe - moderate mobility ($p_{12} = 0.02$, $p_{21} = 0.1$)
- Botswana - high mobility ($p_{12} = 0.02$, $p_{21} = 0.58$)
- Zimbabwe - high mobility ($p_{12} = 0.02$, $p_{21} = 0.58$)

country than in it. This result confirms that if the malaria burden were to get worse in Zimbabwe, achieving overall elimination would prove to be a lot harder. Indeed, it would only be possible with substantial intervention success in Botswana along with a significant increase in mobility from Zimbabwe into Botswana.

4 Discussion

While massive improvements have been made on a global scale in managing malaria, we are still not at the level of elimination. For 2017, the World Health Organization (WHO) estimates over 400,000 deaths to malaria, the vast majority of which occurred in 17 countries [57]. Significant challenges remain in the management of malaria, including climate change and emerging resistance of mosquitoes to insecticides [25, 42]. Further complicating elimination efforts, recent data suggests that malaria incidence is actually increasing in multiple countries that were previously on positive trajectories [50, 57]. As such, this work, which considers the dynamics and management of malaria in multiple connected countries, is particularly timely.

Here, we considered the dynamics of a vector and human population in two patches that represent Botswana and Zimbabwe. We focused on how strategies and treatments in one country are impacted by the other country. Our elasticity analysis and simulations demonstrated that elimination is most easily attained when countries work together. We considered the impact of different intervention strategies by varying parameters in each patch independently. Finally, we show that, since Zimbabwe has a much larger human population with a higher R_0, it can significantly influence the efforts in Botswana. Similar source-sink dynamics have been shown to be important in Madagascar [40].

We consider the impact of migration on both the R_0 and total number of infections. These measures provide complementary information about disease transmission dynamics. R_0 is a simple way to interpret the long-term behavior as either increasing or decreasing disease spread. In contrast, the number of yearly cases provides immediate feedback that is comparable to reported clinical incidence data, and perhaps is more informative for economic and clinical planning of interventions currently. In particular, the R_0 can be less than one, suggesting transmission is decreasing and eventually would be eliminated, while yearly cases are still nonzero, perhaps still advocating for intervention to decrease the disease load in the population.

To facilitate analysis, our work considered a simplified model of malaria dynamics. We now note two features that we did not include and would have the potential to impact our findings. First, we followed a previous approach to modeling human mobility which considers visitation between patches. We note that this allows humans to be infected in either patch. That is, we assume mosquitoes do not move and only infect humans within their patch. While this assumption is likely to make sense for short term visitation, this has created the effect in our model where increasing the amount of time an individual in Zimbabwe spends in Botswana does not change the incidence of malaria in the vector population. Indeed, empirical evidence suggests that mosquitoes can move long distances when winds are high [39]. Second, our model does not consider death of the human population. It has been previously observed that such features can introduce bifurcations which fundamentally alter the system dynamics [4, 18, 32, 45].

We theoretically explored the role of movement by considering the p matrix over a range of values. One could instead use empirical movement data, such as tracking visas, surveys, or potentially genetics, to estimate p [49, 71]. For example, in [40], the authors use mobile phone data to quantify movement and link to clinical cases. Their results were qualitatively similar, with movement from highland sources maintaining transmission loads in lowland sinks.

Because malaria elimination remains an important problem, mathematical modeling will continue to be a powerful tool for evaluating treatment strategies and generating predictions. The modeling framework we have chosen may be easily generalized. First, we note that humans in our model have a home patch. As such our human mobility is that of short visitation (Lagrangian dynamics) rather than migration (Eulerian dynamics) [23, 62, 63]. Our model could be adapted to include both types of mobility. Second, our model framework can clearly include multiple patches. Because many of the countries with the highest malaria incidence are geographically adjacent, it is clear that to fully evaluate elimination strategies multiple countries must be simultaneously depicted. Third, in our model the total number of vectors and humans remains constant. This allowed us to only model the fraction of infected populations in each category. However, an alternate approach which would allow the total populations to change would be to separately model the susceptible and infected populations in each category as was done recently in [12]. Fourth, as has been noted in many recent studies, global climate change will significantly impact vector populations and for longer term elimination evaluation such effects should be included [28, 42]. Finally, mathematical models such as ours require tuning of parameters. The process of linking empirical observations to parameters is complicated. While our metric of the number of new infections provides an easier way to compare model output to data (for example, WHO data which reports the number of new malaria cases), fitting the model to data remains a challenge. In our work, some parameters come from the literature while some are fit under the assumption that the mean Malaria Atlas Project reported R_0 values for each country were correct. However, this led to predictions in new cases that were far greater than the WHO reported cases in each country. Therefore, in the future, more care needs to be taken when parameterizing the model and making sure it is consistent with the WHO reported cases in each country. In addition, as mosquito populations evolve resistance to insecticides it is possible that to fully capture their behavior, such factors need to be included [35].

As our work has shown, malaria elimination will require the concerted effort across geopolitical boundaries. Mathematical modeling will be a powerful tool for evaluating intervention strategies and directing resources. Malaria elimination is an important human health goal and requires interactions between health organizations, scientists, and governments [31].

Acknowledgments The work described herein was initiated during the Collaborative Workshop for Women in Mathematical Biology hosted by the Institute for Pure and Applied Mathematics at the University of California, Los Angeles in June 2019. Funding for the workshop was provided by IPAM, the Association for Women in Mathematics' NSF ADVANCE "Career Advancement for Women Through Research-Focused Networks" (NSF-HRD 1500481) and the Society for Industrial and Applied Mathematics. The authors thank the organizers of the IPAM-WBIO workshop (Rebecca Segal, Blerta Shtylla, and Suzanne Sindi) for facilitating this research.

Additionally, AG acknowledges support by NIH R35GM133481 and SS acknowledges support from the Joint DMS/NIGMS Initiative to Support Research at the Interface of the Biological and Mathematical Sciences (R01-GM126548).

References

1. Miguel A Acevedo, Olivia Prosper, Kenneth Lopiano, Nick Ruktanonchai, T Trevor Caughlin, Maia Martcheva, Craig W Osenberg, and David L Smith. Spatial heterogeneity, host movement and mosquito-borne disease transmission. *PloS one*, 10(6):e0127552, 2015.
2. EA Afari, M Appawu, S Dunyo, A Baffoe-Wilmot, and FK Nkrumah. Malaria infection, morbidity and transmission in two ecological zones southern ghana. *African journal of health sciences*, 2(2):312–315, 1995.
3. Folashade B Agusto. Malaria drug resistance: The impact of human movement and spatial heterogeneity. *Bulletin of mathematical biology*, 76(7):1607–1641, 2014.
4. Folashade B Agusto, Sara Y Del Valle, Kbenesh W Blayneh, Calistus N Ngonghala, Maria J Goncalves, Nianpeng Li, Ruijun Zhao, and Hongfei Gong. The impact of bed-net use on malaria prevalence. *Journal of theoretical biology*, 320:58–65, 2013.

5. Linda JS Allen, Benjamin M Bolker, Yuan Lou, and Andrew L Nevai. Asymptotic profiles of the steady states for an sis epidemic patch model. *SIAM Journal on Applied Mathematics*, 67(5):1283–1309, 2007.

6. Julien Arino, Jonathan R Davis, David Hartley, Richard Jordan, Joy M Miller, and P Van Den Driessche. A multi-species epidemic model with spatial dynamics. *Mathematical Medicine and Biology*, 22(2):129–142, 2005.

7. Julien Arino and P Van den Driessche. A multi-city epidemic model. *Mathematical Population Studies*, 10(3):175–193, 2003.

8. Malaria Map Atlas. Reproductive number under control, 2000-2016, 2019.

9. Lindsay M Beck-Johnson, William A Nelson, Krijn P Paaijmans, Andrew F Read, Matthew B Thomas, and Ottar N Bjørnstad. The importance of temperature fluctuations in understanding mosquito population dynamics and malaria risk. *Royal Society open science*, 4(3):160969, 2017.

10. A Bekessy, L Molineaux, and J Storey. Estimation of incidence and recovery rates of plasmodium falciparum parasitaemia from longitudinal data. *Bulletin of the World Health Organization*, 54(6):685, 1976.

11. Samir Bhatt, DJ Weiss, E Cameron, D Bisanzio, B Mappin, U Dalrymple, KE Battle, CL Moyes, A Henry, PA Eckhoff, et al. The effect of malaria control on plasmodium falciparum in africa between 2000 and 2015. *Nature*, 526(7572):207, 2015.

12. Derdei Bichara and Carlos Castillo-Chavez. Vector-borne diseases models with residence times–a lagrangian perspective. *Mathematical biosciences*, 281:128–138, 2016.

13. Sarah Bonnet, Clement Gouagna, Innocent Safeukui, Jean-Yves Meunier, and Christian Boudin. Comparison of artificial membrane feeding with direct skin feeding to estimate infectiousness of plasmodium falciparum gametocyte carriers to mosquitoes. *Transactions of the Royal Society of tropical Medicine and Hygiene*, 94(1):103–106, 2000.

14. Statistics Botswana. Tourism statistics: Annual report 2017. http://www.statsbots.org.bw/sites/default/files/publications/. Accessed: 2019-07-19.

15. Statistics Botswana. Work permits holders fourth quarter 2018. http://www.statsbots.org.bw/sites/default/files/publications/Work. Accessed: 2019-08-01.

16. Eugene Campbell and Jonathan Crush. Unfriendly neighbours: Contemporary migration from zimbabwe to botswana. 2012.

17. Simon Chihanga, Allison Tatarsky, HT Masendu, D Ntebela, Tjantilili Mosweunyane, Mpho Motlaleng, Godira Segoea, Justin M Cohen, Mercy Puso, Bosiela Segogo, et al. Improving llin utilization and coverage through an innovative distribution and malaria education model: a pilot study in okavango sub-district, botswana. *Malaria journal*, 11(S1):P95, 2012.

18. J. M. Chitnis, Nakul Cushing and J. M. Hyman. Bifurcation analysis of a mathematical model for malaria transmission. *SIAM Journal of Applied Mathematics*, 67(1):24–45, 2006.

19. Nakul Chitnis, James M Hyman, and Jim M Cushing. Determining important parameters in the spread of malaria through the sensitivity analysis of a mathematical model. *Bulletin of mathematical biology*, 70(5):1272, 2008.

20. Thomas S Churcher, Robert E Sinden, Nick J Edwards, Ian D Poulton, Thomas W Rampling, Patrick M Brock, Jamie T Griffin, Leanna M Upton, Sara E Zakutansky, Katarzyna A Sala, et al. Probability of transmission of malaria from mosquito to human is regulated by mosquito parasite density in naive and vaccinated hosts. *PLoS pathogens*, 13(1):e1006108, 2017.

21. Thomas S Churcher, Jean-Francois Trape, and Anna Cohuet. Human-to-mosquito transmission efficiency increases as malaria is controlled. *Nature communications*, 6:6054, 2015.

22. William E Collins and Geoffrey M Jeffery. A retrospective examination of mosquito infection on humans infected with plasmodium falciparum. *The American journal of tropical medicine and hygiene*, 68(3):366–371, 2003.

23. Chris Cosner, John C Beier, Robert Stephen Cantrell, D Impoinvil, Lev Kapitanski, Matthew David Potts, A Troyo, and Shigui Ruan. The effects of human movement on the persistence of vector-borne diseases. *Journal of theoretical biology*, 258(4):550–560, 2009.

24. Hans de Kroon, Anton Plaisier, Jan van Groenendael, and Hal Caswell. Elasticity: the relative contribution of demographic parameters to population growth rate. *Ecology*, 67(5):1427–1431, 1986.
25. Sunil Dhiman. Are malaria elimination efforts on right track? an analysis of gains achieved and challenges ahead. *Infectious diseases of poverty*, 8(1):14, 2019.
26. Thanate Dhirasakdanon, Horst R Thieme, and P Van Den Driessche. A sharp threshold for disease persistence in host metapopulations. *Journal of biological dynamics*, 1(4):363–378, 2007.
27. Christopher Dye and Günther Hasibeder. Population dynamics of mosquito-borne disease: effects of flies which bite some people more frequently than others. *Transactions of the Royal Society of Tropical Medicine and Hygiene*, 80(1):69–77, 1986.
28. Steffen E Eikenberry and Abba B Gumel. Mathematical modeling of climate change and malaria transmission dynamics: a historical review. *Journal of mathematical biology*, 77(4):857–933, 2018.
29. Marisa C Eisenberg, Zhisheng Shuai, Joseph H Tien, and P Van den Driessche. A cholera model in a patchy environment with water and human movement. *Mathematical Biosciences*, 246(1):105–112, 2013.
30. Elimination8. Botswana: Country Overview. https://malariaelimination8.org/botswana/. Accessed: 2019-09-09.
31. Elimination8. SADC Elimination Eight Initiative Annual Report 2019. https://malariaelimination8.org/wp-content/uploads/2020/04/Elimination_8_Annual_Report_2019.pdf. Accessed: 2019-11-27.
32. Xiaomei Feng, Shigui Ruan, Zhidong Teng, and Kai Wang. Stability and backward bifurcation in a malaria transmission model with applications to the control of malaria in china. *Mathematical biosciences*, 266:52–64, 2015.
33. Georg A Funk, Vincent AA Jansen, Sebastian Bonhoeffer, and Timothy Killingback. Spatial models of virus-immune dynamics. *Journal of theoretical biology*, 233(2):221–236, 2005.
34. Daozhou Gao, Yijun Lou, and Shigui Ruan. A periodic ross-macdonald model in a patchy environment. *Discrete and continuous dynamical systems. Series B*, 19(10):3133, 2014.
35. Markus Gildenhard, Evans K Rono, Assetou Diarra, Anne Boissière, Priscila Bascunan, Paola Carrillo-Bustamante, Djeneba Camara, Hanne Krüger, Modibo Mariko, Ramata Mariko, et al. Mosquito microevolution drives plasmodium falciparum dynamics. *Nature microbiology*, 4(6):941, 2019.
36. Weidong Gu, Gerry F Killeen, Charles M Mbogo, James L Regens, John I Githure, and John C Beier. An individual-based model of plasmodium falciparum malaria transmission on the coast of kenya. *Transactions of the Royal Society of Tropical Medicine and Hygiene*, 97(1):43–50, 2003.
37. Günther Hasibeder and Christopher Dye. Population dynamics of mosquito-borne disease: persistence in a completely heterogeneous environment. *Theoretical population biology*, 33(1):31–53, 1988.
38. Ying-Hen Hsieh, P Van den Driessche, and Lin Wang. Impact of travel between patches for spatial spread of disease. *Bulletin of mathematical biology*, 69(4):1355–1375, 2007.
39. Diana L Huestis, Adama Dao, Moussa Diallo, Zana L Sanogo, Djibril Samake, Alpha S Yaro, Yossi Ousman, Yvonne-Marie Linton, Asha Krishna, Laura Veru, et al. Windborne long-distance migration of malaria mosquitoes in the sahel. *Nature*, 574(7778):404–408, 2019.
40. Felana Angella Ihantamalala, Vincent Herbreteau, Feno MJ Rakotoarimanana, Jean Marius Rakotondramanga, Simon Cauchemez, Bienvenue Rahoilijaona, Gwenaëlle Pennober, Caroline O Buckee, Christophe Rogier, Charlotte Jessica Eland Metcalf, et al. Estimating sources and sinks of malaria parasites in madagascar. *Nature communications*, 9(1):1–8, 2018.
41. Emil Ivan, Nigel J. Crowther, Eugene Mutimura, Lawrence Obado Osuwat, Saskia Janssen, and Martin P. Grobusch. Helminthic infections rates and malaria in hiv-infected pregnant women on anti-retroviral therapy in rwanda. *PLOS Neglected Tropical Diseases*, 7(8):1–9, 08 2013.

42. Francois M Moukam Kakmeni, Ritter YA Guimapi, Frank T Ndjomatchoua, Sansoa A Pedro, James Mutunga, and Henri EZ Tonnang. Spatial panorama of malaria prevalence in africa under climate change and interventions scenarios. *International journal of health geographics*, 17(1):2, 2018.

43. Vissundara M Karunasena, Manonath Marasinghe, Carmen Koo, Saliya Amarasinghe, Arundika S Senaratne, Rasika Hasantha, Mihirini Hewavitharana, Hapuarachchige C Hapuarachchi, Hema DB Herath, Rajitha Wickremasinghe, et al. The first introduced malaria case reported from sri lanka after elimination: implications for preventing the re-introduction of malaria in recently eliminated countries. *Malaria journal*, 18(1):210, 2019.

44. Dominik Kopiński and Andrzej Polus. Is Botswana creating a new Gaza strip? an analysis of the 'fence discourse'. *Crossing African Borders: Migration and Mobility*, page 98, 2017.

45. Guihua Li and Zhen Jin. Bifurcation analysis in models for vector-borne diseases with logistic growth. *The Scientific World Journal*, 2014, 2014.

46. Michael Y Li and Zhisheng Shuai. Global stability of an epidemic model in a patchy environment. *Canadian Applied Mathematics Quarterly*, 17(1):175–187, 2009.

47. Rongsong Liu, Jiangping Shuai, Jianhong Wu, and Huaiping Zhu. Modeling spatial spread of west nile virus and impact of directional dispersal of birds. *Mathematical Biosciences & Engineering*, 3(1):145, 2006.

48. Sandip Mandal, Ram Rup Sarkar, and Somdatta Sinha. Mathematical models of malaria-a review. *Malaria journal*, 10(1):202, 2011.

49. John M Marshall, Sean L Wu, Samson S Kiware, Micky Ndhlovu, André Lin Ouédraogo, Mahamoudou B Touré, Hugh J Sturrock, Azra C Ghani, Neil M Ferguson, et al. Mathematical models of human mobility of relevance to malaria transmission in africa. *Scientific reports*, 8(1):1–12, 2018.

50. Amy Maxmen. How to defuse malaria's ticking time bomb. *Nature*, 559:458–465, 2018.

51. K Moakofhi, JK Edwards, M Motlaleng, J Namboze, W Butt, M Obopile, T Mosweunyane, M Manzi, KC Takarinda, and P Owiti. Advances in malaria elimination in botswana: a dramatic shift to parasitological diagnosis, 2008–2014. *Public health action*, 8(1):S34–S38, 2018.

52. Louis Molineaux, Gabriele Gramiccia, World Health Organization, et al. The garki project: research on the epidemiology and control of malaria in the sudan savanna of west africa, 1980.

53. M Motlaleng, J Edwards, J Namboze, W Butt, K Moakofhi, M Obopile, M Manzi, KC Takarinda, R Zachariah, P Owiti, et al. Driving towards malaria elimination in botswana by 2018: progress on case-based surveillance, 2013–2014. *Public health action*, 8(1):S24–S28, 2018.

54. Johanna R Ohm, Francesco Baldini, Priscille Barreaux, Thierry Lefevre, Penelope A Lynch, Eunho Suh, Shelley A Whitehead, and Matthew B Thomas. Rethinking the extrinsic incubation period of malaria parasites. *Parasites & vectors*, 11(1):178, 2018.

55. John Okano. Personal Communication, 2019.

56. Justin T Okano, Katie Sharp, Eugenio Valdano, Laurence Palk, and Sally Blower. Hiv transmission and source–sink dynamics in sub-saharan africa. *The Lancet HIV*, 2020.

57. World Health Organization. WHO World Malaria Report 2018. https://apps.who.int/iris/bitstream/handle/10665/275867/9789241565653-eng.pdf?ua=1. Accessed: 2019-06-18.

58. Panagiotis Pergantas, Andreas Tsatsaris, Chrisovalantis Malesios, Georgia Kriparakou, Nikolaos Demiris, and Yiannis Tselentis. A spatial predictive model for malaria resurgence in central greece integrating entomological, environmental and social data. *PLOS ONE*, 12(6):1–15, 06 2017.

59. Olivia Prosper, Nick Ruktanonchai, and Maia Martcheva. Assessing the role of spatial heterogeneity and human movement in malaria dynamics and control. *Journal of Theoretical Biology*, 303:1–14, 2012.

60. Diego J Rodríguez and Lourdes Torres-Sorando. Models of infectious diseases in spatially heterogeneous environments. *Bulletin of Mathematical Biology*, 63(3):547–571, 2001.

61. Shigui Ruan, Wendi Wang, and Simon A Levin. The effect of global travel on the spread of sars. *Mathematical Biosciences & Engineering*, 3(1):205, 2006.

62. Nick W Ruktanonchai, Patrick DeLeenheer, Andrew J Tatem, Victor A Alegana, T Trevor Caughlin, Elisabeth zu Erbach-Schoenberg, Christopher Lourenço, Corrine W Ruktanonchai, and David L Smith. Identifying malaria transmission foci for elimination using human mobility data. *PLoS computational biology*, 12(4):e1004846, 2016.

63. Nick W Ruktanonchai, David L Smith, and Patrick De Leenheer. Parasite sources and sinks in a patched ross–macdonald malaria model with human and mosquito movement: implications for control. *Mathematical biosciences*, 279:90–101, 2016.

64. Shadreck Sande, Moses Zimba, Joseph Mberikunashe, Andrew Tangwena, and Anderson Chimusoro. Progress towards malaria elimination in zimbabwe with special reference to the period 2003–2015. *Malaria journal*, 16(1):295, 2017.

65. Chihanga Simon, Kentse Moakofhi, Tjantilili Mosweunyane, Haruna Baba Jibril, Bornapate Nkomo, Mpho Motlaleng, Davies Sedisa Ntebela, Emmanuel Chanda, and Ubydul Haque. Malaria control in botswana, 2008–2012: the path towards elimination. *Malaria journal*, 12(1):458, 2013.

66. David L Smith, Katherine E Battle, Simon I Hay, Christopher M Barker, Thomas W Scott, and F Ellis McKenzie. Ross, macdonald, and a theory for the dynamics and control of mosquito-transmitted pathogens. *PLoS pathogens*, 8(4):e1002588, 2012.

67. David L Smith and F Ellis McKenzie. Statics and dynamics of malaria infection in anopheles mosquitoes. *Malaria journal*, 3(1):13, 2004.

68. Gonzalo P Suarez, Oyita Udiani, Brian F Allan, Candice Price, Sadie J Ryan, Eric Lofgren, Alin Coman, Chris M Stone, Lazaros K Gallos, and Nina H Fefferman. A generic arboviral model framework for exploring trade-offs between vector control and environmental concerns. *Journal of Theoretical Biology*, page 110161, 2020.

69. Mohammad Suleman. Malaria in afghan refugees in pakistan. *Transactions of The Royal Society of Tropical Medicine and Hygiene*, 82(1):44–47, 1998.

70. Oscar Tapera. Determinants of long-lasting insecticidal net ownership and utilization in malaria transmission regions: evidence from zimbabwe demographic and health surveys. *Malaria journal*, 18(1):1–7, 2019.

71. Sofonias Tessema, Amy Wesolowski, Anna Chen, Maxwell Murphy, Jordan Wilheim, Anna-Rosa Mupiri, Nick W Ruktanonchai, Victor A Alegana, Andrew J Tatem, Munyaradzi Tambo, et al. Using parasite genetic and human mobility data to infer local and cross-border malaria connectivity in southern africa. *Elife*, 8:e43510, 2019.

72. RBM Partnership to End Malaria. Malaria Strategic Plan – 2010-2015: Towards Malaria Elimination. https://endmalaria.org/sites/default/files/botswa2010-2015.pdf. Accessed: 2019-06-18.

73. United Nations Children's Fund (UNICEF). Distributing long lasting insecticide treated nets. https://www.unicef.org/cbsc/index_55833.html. Accessed: 2019-09-3.

74. Pauline Van den Driessche and James Watmough. Reproduction numbers and sub-threshold endemic equilibria for compartmental models of disease transmission. *Mathematical biosciences*, 180(1-2):29–48, 2002.

75. Wendi Wang and Xiao-Qiang Zhao. An epidemic model in a patchy environment. *Mathematical biosciences*, 190(1):97–112, 2004.

76. Wendi Wang and Xiao-Qiang Zhao. An age-structured epidemic model in a patchy environment. *SIAM Journal on Applied Mathematics*, 65(5):1597–1614, 2005.

77. Peter Winskill, Patrick G Walker, Richard E Cibulskis, and Azra C Ghani. Prioritizing the scale-up of interventions for malaria control and elimination. *Malaria journal*, 18(1):122, 2019.

78. Hyun M Yang and Marcelo U Ferreira. Assessing the effects of global warming and local social and economic conditions on the malaria transmission. *Revista de saude publica*, 34(3):214–222, 2000.

Investigating the Impact of Combination Phage and Antibiotic Therapy: A Modeling Study

Selenne Banuelos, Hayriye Gulbudak, Mary Ann Horn, Qimin Huang, Aadrita Nandi, Hwayeon Ryu, and Rebecca Segal

Abstract Antimicrobial resistance (AMR) is a serious threat to global health today. The spread of AMR, along with the lack of new drug classes in the antibiotic pipeline, has resulted in a renewed interest in phage therapy, which is the use of

The original version of this chapter was revised: Affiliations of the authors "Mary Ann Horn" and "Qimin Huang" have been corrected. A correction to this chapter is available at https://doi.org/10.1007/978-3-030-57129-0_9

Authors Selenne Banuelos, Hayriye Gulbudak, Mary Ann Horn, Qimin Huang, Aadrita Nandi, Hwayeon Ryu, and Rebecca Segal have equally contributed to this chapter.

S. Banuelos
Department of Mathematics, California State University Channel Islands, Camarillo, CA, USA
e-mail: selenne.banuelos@csuci.edu

H. Gulbudak
Department of Mathematics, University of Louisiana at Lafayette, Lafayette, LA, USA
e-mail: hayriye.gulbudak@louisiana.edu

M. A. Horn · Q. Huang (✉)
Department of Mathematics, Applied Mathematics, and Statistics, Case Western Reserve University, Cleveland, OH, USA
e-mail: maryann.horn@case.edu; qxh119@case.edu

A. Nandi
Institute for Global Health, Feinberg School of Medicine, Northwestern University, Chicago, IL, USA
e-mail: aadrita.nandi@northwestern.edu

H. Ryu
Department of Mathematics and Statistics, Elon University, Elon, NC, USA
e-mail: hryu@elon.edu

R. Segal
Department of Mathematics and Applied Mathematics, Virginia Commonwealth University, Richmond, VA, USA
e-mail: rasegal@vcu.edu

© The Association for Women in Mathematics and the Author(s) 2021, corrected publication 2021
R. Segal et al. (eds.), *Using Mathematics to Understand Biological Complexity*, Association for Women in Mathematics Series 22, https://doi.org/10.1007/978-3-030-57129-0_6

bacteriophages to treat pathogenic bacterial infections. This therapy, which was successfully used to treat a variety of infections in the early twentieth century, had been largely dismissed due to the discovery of easy to use antibiotics. However, the continuing emergence of antibiotic resistance has motivated new interest in the use of phage therapy to treat bacterial infections. Though various models have been developed to address the AMR-related issues, there are very few studies that consider the effect of phage-antibiotic combination therapy. Moreover, some biological details such as the effect of the immune system on phage have been neglected. To address these limitations, we utilized a mathematical model to examine the role of the immune response in concert with phage-antibiotic combination therapy compounded with the effects of the immune system on the phages being used for treatment. We explore the effect of phage-antibiotic combination therapy by adjusting the phage and antibiotics dose or altering the timing. The model results show that it is important to consider the host immune system in the model and that frequency and dose of treatment are important considerations for the effectiveness of treatment. Our study can lead to development of optimal antibiotic use and further reduce the health risks of the human-animal-plant-ecosystem interface caused by AMR.

Keywords Mathematical modeling · Phage therapy · Antibiotics

1 Introduction

Antimicrobial resistance (AMR) is a serious threat to global health. The Centers for Disease Control and Prevention (CDC) estimates that at least 2 million people become infected by antibiotic-resistant bacteria and at least 23,000 people die each year as a direct result of these infections, costing the United States $55 billion annually [3]. Infections caused by bacteria are usually treated with antibiotics. However, due to over-prescribing and mis-prescribing, many strains of bacteria have become resistant to currently available antibiotics. A list of antibiotic-resistant pathogens, a catalog of 12 families of bacteria for which new antibiotics are urgently needed, has been generated by the World Health Organization (WHO) [22]. Nevertheless, since bacteria evolve resistance to antibiotics at a relatively rapid rate, there has been less commercial interest in developing new antibiotics. Only 6 new antibiotics were approved by the Food and Drug Administration (FDA) for use in the United States from 2010 to 2016, an obvious downward trend compared to the 16 new antibiotics approved by FDA between 1983 and 1987 [41]. In 2015, a global action plan on antimicrobial resistance (GAP-AMR) was endorsed at the World Health Assembly, and one of the five strategic objectives of the GAP-AMR is to optimize the use of antimicrobial agents [65]. In 2018, the U.S. government launched the Antimicrobial Resistance Challenge to call for leaders from around the world to work together to improve antibiotic use, accelerate research on new antibiotics and antibiotic alternatives [3]. The spread of antimicrobial resistance combined with the lack of new drug classes in the antibiotic pipeline has resulted in a resurgence of interest in phage therapy.

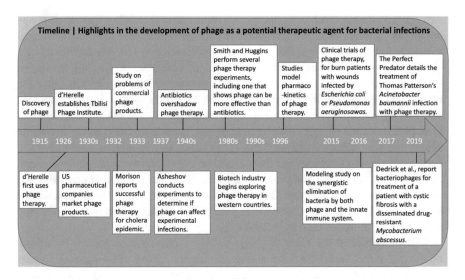

Fig. 1 A timeline of important events in the history of phage therapy. (Adapted and updated from [37])

Phage therapy is the use of bacteriophages to treat pathogenic bacterial infections. Before the widespread use of antibiotics, phage therapy was successfully applied in treating a variety of infections in the 1920s and 1930s [45]. Due to a poor understanding of the biological nature of phages, medical limitations of the day, and introduction of broader spectrum antibiotics, phage therapy was largely dismissed by most of western medicine in the 1940s [39]. However, the rise of antibiotic resistance has resulted in renewed interest in using phage to treat bacterial infections [53]. One of the first international, single-blind clinical trials of phage therapy, which aimed to target 220 burn patients with wounds infected by *Escherichia coli* or *Pseudomonas aeruginosa*, was launched in 2015 [17, 33, 47]. Furthermore, clinical trials are currently underway to explore phage treatment for infections caused by *Staphylococcus aureus*, particularly for respiratory tract infection (e.g., pneumonia), and to reduce the population of pathogens in ready-to-eat foods and meat [1, 25–27, 34, 51]. In contrast to antibiotics, bacteria sensitivity to phages is largely specific for both species and strain, which can be considered as a major advantage, since the effects of antibiotics on commensal gut microbes are notorious for secondary outcomes such as antibiotic-associated diarrhea and *C. difficile* infection [46]. See Fig. 1 for a timeline of important events in the development and use of phage therapy.

Because the problem of antibiotic resistant bacteria is complex and growing, with no known solution, various mathematical models have been proposed to explore the dynamics of the variety of systems involved. Most models focus on the transmission dynamics of antibiotic-resistant bacteria at the host population level [4, 5, 7–11, 14, 18–21, 28, 30–32, 40, 58–63]; some focus on exploring the relative contributions of antibiotics and immune response in the treatment of infection on the bacterial population level [2, 6, 36, 38, 52]. Now, with increasing interest in phage therapy as an alternative or supplement to antibiotic treatment [39], mathematical models incorporating phage therapy have been developed [13, 16, 23, 35, 37, 43, 44, 49, 50, 54, 57, 64]. In particular, Rodriguez-Gonzalez et al. [50] developed a mathematical model of phage-antibiotic combination therapy, representing the interactions among bacteria, phage, antibiotics, and the innate immune system, but ignoring the effect of immune system on phages. Some evidence shows that while phages do not trigger an immune response, bacteria-boosted innate immunity activity can act against the phages [29]. This finding may explain instances of phage ineffectiveness and suggests that there could be better protocols for phage therapy. To include this important component, we extended earlier models, in particular, the model developed by Rodriguez-Gonzalez, et al. [50]. The goal is to understand the role of the immune response in concert with phage-antibiotic combination therapy by introducing immune activity related to phages to the model.

We aim to explore the effect of phage-antibiotic combination therapy by adjusting the phage and antibiotic doses and/or altering the timing of the dose(s). Details of the system of nonlinear, ordinary differential equations which take into account the interactions among bacteria, phage, antibiotics, and the immune system are given in Sect. 2. In Sects. 3 and 4, equilibria and sensitivity analysis of some reduced cases are provided. In Sect. 5, the simulation results exploring various infection and treatment scenarios are presented, followed by a discussion in Sect. 6.

2   Mathematical Model

We present a deterministic antibiotic-phage combination therapy model that describes density-dependent interactions between two strains of bacteria, phage, antibiotics, and the host immune response. The model development builds on the work by Leung and Weitz [34], and Rodriguez-Gonzalez, et al. [50]. The model in [34] includes phage-sensitive bacteria, phage, and a saturating innate immune response. The phage therapy model in [50] extended [34] to include two bacteria strains, phage therapy, antibiotic treatment and some immune response components. The model presented here adds biological functions not included in these previous models: interactions of the immune response with phage, and the decay of the immune response. See Fig. 2 for a schematic diagram of our model.

$$\frac{dB_P}{dt} = \overbrace{r_P B_P \left(1 - \frac{B_{tot}}{K_c}\right)(1 - \mu_P)}^{\text{Growth of } B_P} + \overbrace{\mu_A r_A B_A \left(1 - \frac{B_{tot}}{K_c}\right)}^{\text{Growth of } B_A \text{ mutated to } B_P}$$

$$- \overbrace{\frac{\epsilon I B_P}{1 + \frac{B_{tot}}{K_D}}}^{\text{Immune killing}} - \overbrace{B_P F(P)}^{Lysis},$$

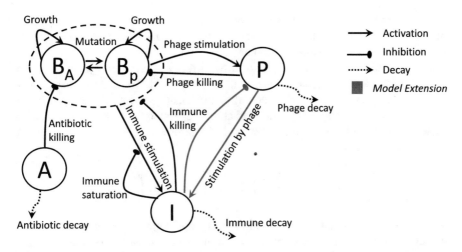

Fig. 2 Schematic diagram of the extended phage-antibiotic combination therapy model. Antibiotic-sensitive bacteria (B_A) and phage-sensitive bacteria (B_P) are targeted by antibiotic (A) and the phage (P), respectively. The immune response interactions with both bacterial strains are included in the model. In addition, our model extension building on the model in [50] includes the innate immunity (I) stimulation by the presence of phage (in red arrows) and the decay of the immune response

$$\frac{dB_A}{dt} = \overbrace{r_A B_A \left(1 - \frac{B_{tot}}{K_c}\right)(1 - \mu_A)}^{\text{Growth of } B_A} + \overbrace{\mu_P r_P B_P \left(1 - \frac{B_{tot}}{K_c}\right)}^{\text{Growth of } B_P \text{ mutated to } B_A}$$
$$- \overbrace{\frac{\epsilon I B_A}{1 + \frac{B_{tot}}{K_D}}}^{\text{Immune killing}} - \overbrace{K_{kill} \frac{A}{EC_{50} + A} B_A}^{\text{Antibiotic killing}},$$

$$\frac{dA}{dt} = A_I - \theta A,$$

$$\frac{dP}{dt} = \overbrace{\tilde{\beta} B_P F(P)}^{\text{Replication}} - \overbrace{\omega P}^{\text{Decay}} - \overbrace{\kappa P I}^{\text{Immune killing}},$$

$$\frac{dI}{dt} = \overbrace{\alpha I \left(1 - \frac{I}{K_I}\right)\left(\frac{B_{tot}}{B_{tot} + K_N}\right)}^{\text{Stimulation response to bacteria}} + \overbrace{\beta I \left(1 - \frac{I}{K_I}\right)\left(\frac{P}{P + K_M}\right)}^{\text{Stimulation response to phage}} - \overbrace{d I}^{\text{Decay}}.$$

The phage-sensitive (antibiotic-resistant) bacteria, denoted as B_P, respond to treatment by phages, P, whereas the antibiotic-sensitive (phage-resistant) bacteria, B_A, respond to treatment by antibiotics, A. It can be assumed that bacteria is either resistant to phages or to antibiotics due to conservation of evolutionary resources in the bacteria [15]. The total immune response (I) is activated by the presence of bacteria and phages, and attacks both bacterial strains.

The bacteria grow logistically with growth rate r_i, carrying capacity K_c, and density dependence $B_{tot} = \sum_i B_i$, where $i \in \{A, P\}$. We assume that phage-sensitive bacteria mutate to become antibiotic sensitive bacteria with probability μ_P. Similarly, μ_A represents the probability of emergence of phage-sensitive mutants from antibiotic-sensitive bacteria. Therefore the growth of the bacteria population is modeled as:

$$\frac{dB_i}{dt} = r_i B_i \left(1 - \frac{B_{tot}}{K_c}\right)(1 - \mu_i) + \mu_j r_j B_j \left(1 - \frac{B_{tot}}{K_c}\right)$$

where $i, j \in \{A, P\}$ and $i \neq j$. As in [50] both populations of bacteria are killed by an activated innate immune response which includes a density-dependent immune evasion by bacteria. That is, the mass action killing term, $\epsilon I B_i$, is scaled by the parameter $\left(1 + \frac{B_{tot}}{K_D}\right)^{-1}$. See [34] for more details. The decrease in density of B_A by the antibiotic treatment is approximated by a Hill function as in [50]. The phage-sensitive bacteria are infected and lysed by phage at a rate of $F(P)$. Following the work in [50] two phage infection modalities, $F(P)$, are considered – homogeneous mixing and heterogeneous mixing. The homogeneous mixing modality is given by $F(P) = \phi P$ so that the infection rate is proportional to the phage density. The second modality is given by $F(P) = \phi P^\gamma$ where γ is the power-law exponent. The homogeneous mixing modality is assumed for our analytical results in Sect. 3 and sensitivity analysis in Sect. 4, whereas the heterogeneous mixing model is used for the numerical analysis in Sect. 5. Mathematical analysis is not valid with fractional exponents, but it is likely that the phage distribution is heterogeneous.

The growth in the phage density is due to the release of phage through lytic infection of B_P at a rate of $\widetilde{\beta}$. Free phage particles decay at a rate ω. One of the novel biological features included in this model is the effect of the immune response on the phage virus. The differences in the effectiveness of phage therapy between in vitro and in vivo suggest that the infected mammalian host's immune response may be responsible for bacterial phage resistance [29]. The per capita kill rate of phage by the immune response is denoted by κ.

As in [34] we assume there is a saturated innate immune response that is activated by bacteria. We have included a saturated innate immune response that is activated by the presence of phage where β is the maximum growth rage, K_I is the maximum capacity, and K_M is the phage concentration at which the immune response growth rate is half its maximum. In addition, we assume that d is the rate of decay in the immune response. We assume that once the antibiotic treatment is administered it is injected at a constant rate where $A^* = A_I/\theta$. Parameter values are given in Table 1.

Table 1 Parameters and Descriptions (*CFU* colony-forming unit, *PFU* plaque-forming unit)

Symbol	Description	Value	Reference
B_P	Density of phage sensitive bacteria	7.4×10^7 CFU/g	[34, 50]
B_A	Density of antibiotic sensitive bacteria	7.4×10^7 CFU/g	[34, 50]
B_{tot}	Total density of bacteria	$B_P + B_A$	[50]
K_c	Bacterial carrying capacity	10^{10} CFU/g	[50]
μ_P	Probability of emergence of antibiotic sensitive mutants per cellular division	2.85×10^8	[42, 50]
μ_A	Probability of emergence of phage sensitive mutants per cellular division	2.85×10^8	[42, 50]
κ	Killing rate of phage by innate immune response	$10^{-3}\,\mathrm{h}^{-1}$	[29]
r_P	Maximum growth rate of phage sensitive bacteria	0.75	[24, 50]
r_A	Maximum growth rate of antibiotic sensitive bacteria	0.675	[50]
K_D	Bacteria concentration at which immune response is half as effective	4.1×10^7 CFU/g	[50, 56]
K_N	Bacteria concentration when immune response growth rate is half its maximum	10^7 CFU/g	[50, 66]
K_I	Maximum capacity of immune response	2.7×10^6 cell/g	[48, 50]
ϵ	B_P and B_A killing rate by immune system	8.2×10^{-8} g/h	[24, 50]
α	Maximum growth rate of immune system	$0.97\,\mathrm{h}^{-1}$	[50]
β	Rate of change in immunity by Phage	$10^{-5}\,\mathrm{h}^{-1}$	Estimated
K_M	Maximum capacity of phage	10^7 PFU/g	Estimated
d	Decay rate of immune response	$10^{-4}\,\mathrm{h}^{-1}$	Estimated
$\widetilde{\beta}$	Burst size of phage	100	[50]
ϕ	Phage absorption rate	$5.4 * 10^{-8}$	[50]
γ	Power law exponent for heterogeneous mixing	0.6	[50]
θ	Antibiotic elimination rate from serum samples	$0.53\,\mathrm{h}^{-1}$	[50]
EC_{50}	Concentration of antibiotic at which the killing rate is half its maximum	0.3697 ug/ml	[50]

3 Analytical Results

Here, we analytically explore possible treatment outcomes via equilibria analysis.

In Table 2 we summarize possible equilibria of the system, suggesting that infection dynamics can result in any of the following cases under combination treatment:

I. Combination treatment fails.
II. Partial success is gained, since antibiotic sensitive bacteria die out as a result of combination (or drug-only treatment). Yet, the equilibria analysis, detailed below, suggests that there might be up to three outcomes, indicating that the system might have bistable dynamics; i.e., the treatment outcomes might depend on the initial bacteria density, treatment doses and timing (see numerical results section).

Table 2 Possible equilibria of the system

Case	Description	Equilibrium $\mathcal{E}_p^\dagger = (B_P^\dagger, B_A^\dagger, P^\dagger, I^\dagger, A^\dagger)$
(I)	Infection equilibrium	$(*,*,*,*,*)$
(II)	Antibiotic Sensitive Bacteria (ASB)-free equilibrium $(\mathcal{E}_{ASBf})$	$(*,0,*,*,*)$
(III)	Infection-free equilibrium	$(0,0,0,0,*)$
(IV)	Phage-free (Pf) equilibrium $(\mathcal{E}_{Pf})$	$(*,*,0,*,*)$
(V)	Phage & Antibiotic Sensitive Bacteria (ASB)-free equilibrium	$(*,0,0,*,*)$
(VI)	Phage & Phage Sensitive Bacteria (PSB)-free equilibrium	$(0,*,0,*,*)$

III. Successful treatment. Both phage and antibiotic-sensitive bacteria get cleared.
IV. Phage treatment completely fails. It decays before clearing the phage-sensitive bacteria.
V. Phage treatment fails, yet drug treatment successfully eradicates the antibiotic sensitive bacteria.
VI. Drug treatment fails, yet phage therapy eradicates phage sensitive bacteria.

A rigorous mathematical analysis and feasibility of these outcomes require stability analysis. We provide the detailed analysis of some of the cases. Below, we provide the details from the analysis of Case (II). Cases (IV) and (V) are detailed in Appendix. Due to complexity of the system, we explore the possible outcomes derived here using numerical experiments in Sect. 5.

Case II. Antibiotic Sensitive Bacteria (ASB)-free equilibrium $(\mathcal{E}_{ASBf})$. In the absence of antibiotic-sensitive (phage-resistant) bacteria, we obtain the following subsystem:

$$\dot{B}_P = r_p B_P \left(1 - \frac{B_P}{K_c}\right) - \frac{\epsilon I B_P}{1 + \frac{B_P}{K_D}} - B_P F(P), \tag{1}$$

$$\dot{P} = \tilde{\beta} B_P F(P) - \omega P - \kappa P I, \tag{2}$$

$$\dot{I} = \alpha I \left(1 - \frac{I}{K_I}\right)\left(\frac{B_P}{B_P + K_N}\right) + \beta I \left(1 - \frac{I}{K_I}\right)\left(\frac{P}{P + K_M}\right) - dI, \tag{3}$$

$$\dot{A} = A_I - \theta A \tag{4}$$

where $F(P) = \phi P$ and $B_A = 0$. Equilibria of the system are the time-independent solutions. Here we are interested in phage treatment only, i.e., coexistence equilibrium

$$\mathcal{E}_p^\dagger = (B_P^\dagger, 0, P^\dagger, I^\dagger, A^\dagger).$$

By setting the left hand of the system equal to zero, from the first equation, we obtain,

$$r_p \left(1 - \frac{B_P^\dagger}{K_c}\right) = \frac{\epsilon I^\dagger}{1 + \frac{B_P^\dagger}{K_D}} + \phi P^\dagger, \quad \text{where } (B_P^\dagger := f_1(I^\dagger, P^\dagger)). \tag{5}$$

By the second equation, we also have

$$B_P^\dagger = \frac{\omega + \kappa I^\dagger}{\tilde{\beta}\phi} \quad \text{where } (B_P^\dagger := f_2(I^\dagger)) \tag{6}$$

In addition, by the third equation, we get

$$\alpha \left(1 - \frac{I^\dagger}{K_I}\right)\left(\frac{B_P^\dagger}{B_P^\dagger + K_N}\right) = -\beta \left(1 - \frac{I^\dagger}{K_I}\right)\left(\frac{P^\dagger}{P^\dagger + K_M}\right) + d. \tag{7}$$

Rearranging the equality (7), we have

$$B_P^\dagger = \frac{\alpha \left(1 - \frac{I^\dagger}{K_I}\right) - \left(d - \beta \left(1 - \frac{I^\dagger}{K_I}\right)\left(\frac{P^\dagger}{P^\dagger + K_M}\right)\right)}{K_N \left(d - \beta \left(1 - \frac{I^\dagger}{K_I}\right)\left(\frac{P^\dagger}{P^\dagger + K_M}\right)\right)}, \quad \text{where } (B_P^\dagger := f_3(I^\dagger, P^\dagger)). \tag{8}$$

By the equality (8), we also have

$$P^\dagger = \frac{K_M \left(d \left(K_N B_P^\dagger + 1\right) - \alpha \left(1 - \frac{I^\dagger}{K_I}\right)\right)}{\left(1 - \frac{I^\dagger}{K_I}\right)\left(\beta[B_P^\dagger K_N - 1] + \alpha\right) + d \left(1 - K_N B_P^\dagger\right)}. \tag{9}$$

Substituting (9) into (5), we get

$$\frac{r_p}{\phi}\left(1 - \frac{B_P^\dagger}{K_c}\right) - \frac{\epsilon I^\dagger}{\phi(1 + \frac{B_P^\dagger}{K_D})} = \frac{K_M \left(d \left(K_N B_P^\dagger + 1\right) - \alpha \left(1 - \frac{I^\dagger}{K_I}\right)\right)}{\left(1 - \frac{I^\dagger}{K_I}\right)\left(\beta[B_P^\dagger K_N - 1] + \alpha\right) + d \left(1 - K_N B_P^\dagger\right)}, \tag{10}$$

where $(B_P^\dagger := f_4(I^\dagger))$.

Finally, by substituting the right hand side of the equation (6) into (10), we get the following equality as a function of immune equilibrium component, $I^\dagger$:

$$\underbrace{\frac{r_p}{\phi}\left(1 - \frac{\omega + \kappa I^\dagger}{\tilde{\beta}\phi K_c}\right) - \frac{\epsilon I^\dagger}{\phi(1 + \frac{\omega + \kappa I^\dagger}{\tilde{\beta}\phi K_D})}}_{h(I^\dagger)}$$

$$= \frac{K_M \left(d \left(K_N \frac{\omega + \kappa I^\dagger}{\tilde{\beta} \phi} + 1 \right) - \alpha \left(1 - \frac{I^\dagger}{K_I} \right) \right)}{\underbrace{\left(1 - \frac{I^\dagger}{K_I} \right) \left(\beta [\frac{\omega + \kappa I^\dagger}{\tilde{\beta} \phi} K_N - 1] + \alpha \right) + d \left(1 - K_N \frac{\omega + \kappa I^\dagger}{\tilde{\beta} \phi} \right)}_{z(I^\dagger)}}. \tag{11}$$

The positive intersections of the functions $h(I^\dagger)$, and $z(I^\dagger)$ provide the possible immune equilibrium component, $I^\dagger$. Notice that the left hand side of the equality, $h(I^\dagger)$, is a decreasing function of $I^\dagger$. Moreover, the function $z(I^\dagger)$ has a unique zero:

$$I_0 = \frac{\alpha - d \left(1 + \frac{K_N w}{\tilde{\beta} \phi} \right)}{\frac{d K_N \kappa}{\tilde{\beta} \phi} + \frac{\alpha}{K_I}},$$

and two asymptotes $I_{1,2}$:

$$I_{1,2} = \frac{-a_2 \pm \sqrt{a_2^2 - 4 a_1 a_3}}{2 a_1},$$

where

$$a_1 = -\frac{\beta \kappa K_N}{\tilde{\beta} \phi K_I},$$

$$a_2 = -\frac{1}{K_I} \left(\frac{\beta w}{\tilde{\beta} \phi} - \beta + \alpha \right) + \frac{\beta \kappa K_N}{\tilde{\beta} \phi} - \frac{d K_N \kappa}{\tilde{\beta} \phi},$$

$$a_3 = \frac{\beta w}{\tilde{\beta} \phi} - \beta + \alpha + d - \frac{d K_N \omega}{\tilde{\beta} \phi}.$$

Under distinct cases with respect to sign and order of the critical points $I_{0,1,2}$, the subsystem (1) might have zero or up to three possible positive equilibria. Note that whenever $I^\dagger > 0$, we have $B_P^\dagger > 0$ by the equation (6). Therefore we are looking for immune equilibrium component, $I^\dagger : I^\dagger > 0 \Rightarrow P^\dagger > 0$. This result indicates that the system might have bistable dynamics; i.e., the treatment outcomes might depend on the initial bacteria/phage density, treatment doses and timing (see Sect. 5).

4   Sensitivity Analysis

Building on the work in [50], we adopt many parameter values from literature estimates and behavior fitting. However, several of our parameter values are not experimentally measurable. To determine the relative effect of fluctuations in parameter values on the model output, we use Matlab and Simbiology to implement the model and run a sensitivity analysis (similar to the process described in [55]). Sensitivity analysis of parameters for our model will inform us about changes to which parameters would have the most affect on the model transients. The following general steps were performed to produce a global sensitivity analysis for all the parameters over the simulation time period.

First, we established a set of reasonable parameters. The model needs to start at an admissible point in parameter space. We then used this fitted model to generate the discretized sensitivity matrix S. We then used S to rank parameters by sensitivity and set a threshold such that parameters with sensitivity below the threshold (insensitive) are fixed and parameters with sensitivity above the threshold (sensitive) are explored.

To apply this process to our model, we used the referenced values as a starting point as listed in Table 2. Most of these parameter values were used in [50] and we estimated the new parameter values for the full model to achieved biologically reasonable transient output for the model. All four observable model outputs (B_A, B_P, P, and I) were sampled at 10 time points (16, 20, 24, 40, 48 h, and days 3–7). Given that there are 18 model parameters explored, a 40×18 discretized sensitivity matrix S is produced.

Next, we ranked the impact of each parameter on all four observable model outputs (B_A, B_P, P, and I) by calculating a root mean square sensitivity measure, as defined in Brun et al. [12]. For each column j of the normalized sensitivity matrix, we get

$$RMS_j = \sqrt{\frac{1}{n}\sum_{i=1}^{n}\left(\frac{p_j}{y_i}\frac{\partial y_i}{\partial p_j}\right)^2}.$$

Parameter j is deemed insensitive if RMS_j is less than 5% of the value of the maximum RMS value calculated over all parameters. By this measure, 12 parameters were deemed insensitive, as shown in Fig. 3, and fixed at their nominal values in later investigations.

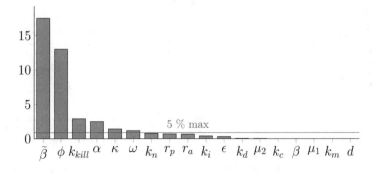

Fig. 3 Relative sensitivities. Values below 5% of the maximum sensitivity value (indicated with the dashed line) are considered insensitive

The model was most sensitive to $\tilde{\beta}$ and ϕ. These modulate the rate of phage replication in the phage equation and also the burst rate of phage infected bacteria. Since $\tilde{\beta}$ only appears in the P equation and it actually multiplies ϕ, in Sect. 5 we explored the effect of ϕ on the model outcome. Although not deemed as sensitive, we also chose to explore the effect of κ, the rate of removal of phage by immune cells, on the model outcomes since it is a new parameter in our extended system.

In the numerical results below, we explore the changes to model transients that result from different choices of these sensitive parameters.

5   Numerical Results

In this section, we explore numerically computed transients for some biological relevant cases of the system and apply our proposed model to investigate the interactions between bacteria, phages, antibiotics and the immune system.

5.1   Exploring the Immune Response

Without any treatments (either phage or antibiotic), Fig. 4 shows that the activated immune response can clear bacteria when bacterial densities (cell densities) are low enough; however, when bacterial densities are sufficiently high, the immune system cannot mount a sufficient response to clear the infection. The complicated role of the immune response in therapeutic application of phage and antibiotics are still overgeneralized here and will be further expanded upon in later versions of the model.

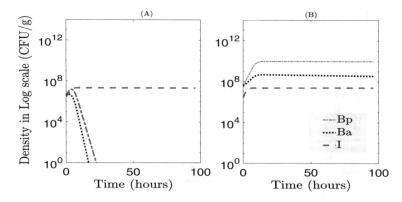

Fig. 4 Model simulations with two levels of initial cell density. (**a**) a low initial cell density $B_P(0) = B_A(0) = 3.7 * 10^7$ CFU/g; (**b**) a high initial cell density $B_P(0) = B_A(0) = 3.7 * 10^8$ CFU/g. No treatments are applied and other parameter values are fixed as in Table 1

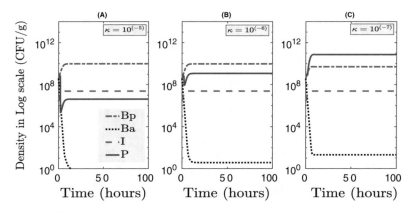

Fig. 5 Model behaviors with different levels of killing rate, κ, of phage by immune cells. (**a**) $\kappa = 10^{-5}$; (**b**) $\kappa = 10^{-6}$; (**c**) $\kappa = 10^{-7}$. Here, $B_P(0) = B_A(0) = 3.7 * 10^8$ CFU/g, both phage and antibiotic treatment were administered 2 h post infection, and other parameter values fixed as in Table 1

In this model, we have included terms to track that immune cell activity in response to the presence of phages during treatment [29]. This is a relevant inclusion to the model because it helps explain instances of phage ineffectiveness. One component of the new model terms, is the parameter κ, which describes the clearance of phages by the host immune system. Because this is a new addition to the model, we investigate the effect the value of κ has on the effectiveness of the combination phage-antibiotic therapy.

In Fig. 5, we use three different values of κ, with other parameter values fixed as in Table 1 and initial bacterial levels $B_P(0) = B_A(0) = 3.7 * 10^8$ CFU/g (colony forming unit per gram). Also during our experiments, both phage and antibiotic therapy are received 2 h post infection ($P = 7.4 * 10^8$ PFU/g, $A = 0.035$ ug/ml) (PFU: plaque forming unit). As can be seen in Fig. 5a–c, our results show that higher values of κ, the killing rate of phages by immune response, results in lower availability of phages at the equilibrium state, but that the final patient outcome is not different.

5.2 Effect of Nonlinear Phage Absorption Rate ϕ

In our simulations, we assume that phage infects and lyses B_P bacteria at a rate $F(P)$, where the function $F(P) = \phi P^{0.6}$ is used to account for heterogeneous mixing. In the above sensitivity analysis (Fig. 3) it is shown that the system's transients are sensitive to the nonlinear phage absorption rate ϕ. We therefore have explored the predicted effectiveness of phage therapy, as it changes with altering ϕ. In Fig. 6, we have shown transients for three different choices of ϕ, with other

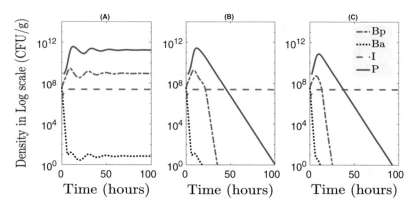

Fig. 6 Model behaviors with different levels of phage absorption rate. (**a**) $\phi = 10.8 * 10^{-8}$; (**b**) $\phi = 13.5 * 10^{-8}$; (**c**) $\phi = 27 * 10^{-8}$. Here, $B_P(0) = B_A(0) = 3.7 * 10^8$ CFU/g, both phages and antibiotic therapy are administered 2 h post infection, and other parameter values are fixed as in Table 1

parameter values fixed in Table 1 and initial bacterial levels $B_P(0) = B_A(0) = 3.7 * 10^8$ CFU/g. In all panels, both phage and antibiotic therapy are received 2 h after the start of the simulation. In Fig. 6a, where phage absorption rate is 2ϕ, the B_A goes to near zero but B_P stays high; while in Fig. 6b, where phage absorption rate is 2.5ϕ, the same initial dose of phages is able to bring down the level of B_P to zero, and we attain the trivial equilibrium; while in Fig. 6c, where phage absorption rate is 5ϕ, the B_P goes to zero and the process occurs faster than compared to Fig. 6b.

5.3 Effect of Time of Administration of Phage Dose

Next, we investigate the effects of timing of phage therapy on the outcome of the infection. In all the simulations, antibiotics are given at the start of the simulation, the initial bacterial levels are $B_P(0) = B_A(0) = 3.7 * 10^7$ CFU/g (a relatively low level), the nonlinear phage absorption rate is 2ϕ, and the other parameter values are fixed as shown in Table 1. In Fig. 7a, the phage dose is given 2 h post infections. We can see that the antibiotic-sensitive bacteria, B_A, decays quickly and goes to equilibrium near zero. Even though phage therapy lowers the B_P bacteria, B_P does not get completely removed from the system and a non-zero equilibrium is achieved for both B_A and B_P. However, in Fig. 7b, when the phage is not administered until 10 h after the start of infection, the density of B_P is already high. This provides more access to bacteria hosts for the phages to use for replication and in turn phages are able to reduce the density of B_P. Then in the absence of B_P, the phage level also goes to zero. These experiments indicates that the timing of phage therapy can be an important factor because phage effectiveness depends on the density of bacteria present in the system.

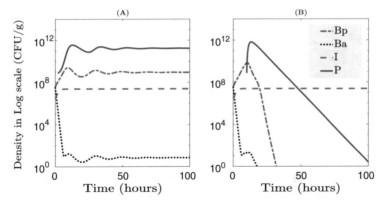

Fig. 7 Model behaviors with different timing of phage therapy. (**a**) Phage dose $7.4 * 10^8$ PFU/g was administered 2 h post infection; (**b**) phage dose $7.4 * 10^8$ PFU/g was administered 10 h post infection. Here, the initial bacterial level is $B_P(0) = B_A(0) = 3.7 * 10^7$ CFU/g (a relatively low level), the nonlinear phage absorption rate is 2ϕ, and the other parameter values are fixed as shown in Table 1

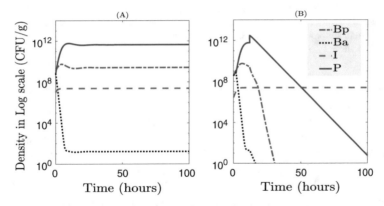

Fig. 8 Model behavior in multi-dose regimen of phage therapy. (**a**) One dose of phage treatment. The only dose $7.4 * 10^8$ PFU/g was administered 2 h post infection; (**b**) two doses of phage treatment. The first dose $7.4 * 10^8$ PFU/g was administered 2 h post infection and the second dose $P = 2.4 * 10^{12}$ PFU/g was administered 10 h after the first dose. Here, the initial bacterial level $B_P(0) = B_A(0) = 3.7 * 10^8$ CFU/g is used, antibiotic therapy is administered 2 h post infection, and parameter values are fixed as shown in Table 1

5.4 Varying Time and Quantity of Phage Dose in Multi-dose Regimen

We continue our experiments by varying the frequency and quantity of the phage therapy dose to explore possible outcomes of phage therapy. For both simulations in Fig. 8 the same initial infection level $B_P(0) = B_A(0) = 3.7 * 10^8$ CFU/g are used, the same antibiotic therapy is administered after 2 h, and the parameter values

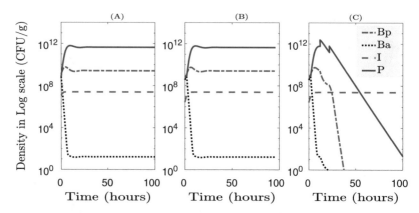

Fig. 9 Model behavior in multi-dose regimen of phage therapy. (**a**) One dose of phage treatment. The only dose $7.4 * 10^8$ PFU/g was administered 2 h post infection; (**b**) two doses of phage treatment. The first dose $7.4 * 10^8$ PFU/g was administered 2 h post infection and the second dose $P = 1.8 * 10^{12}$ PFU/g was administered 10 h after first dose; (**c**) three doses of phage treatment. The first dose $7.4 * 10^8$ PFU/g was administered 2 h post infection, the second dose $P = 1.8 * 10^{12}$ PFU/g was administered 10 h after the first dose, and the third dose $P = 4.5 * 10^{11}$ PFU/g was administered 10 h after the second dose. Here, the initial bacterial level $B_P(0) = B_A(0) = 3.7 * 10^8$ CFU/g is used, antibiotic therapy is administered 2 h post infection, and parameter values are fixed as shown in Table 1

are same as in Table 1. In the first experiment (Fig. 8a), we use only one dose of phage. The dose of phage ($P = 7.4 * 10^8$ PFU/g) is administered 2 h post infection. It is shown that the B_A goes to nearly zero, but B_P goes to a positive equilibrium ($B_P \gg 0$). That is, we do not have a successful treatment. In the second experiment, we explore two doses of phage. As in (Fig. 8a), the first dose ($P = 7.4 * 10^8$ PFU/g) is administered 2 h post infection. Then 10 h after the first dose, the second dose ($P = 2.4 * 10^{12}$ PFU/g) is given. We found that if the amount of second dose of phage is high enough, then the density of B_P goes to zero rapidly, and we obtain a successful treatment at the end. Otherwise, you need to do more doses of phage treatment (See Fig. 9).

In Fig. 9, three experiments are shown. In all simulations, the same initial infection level $B_P(0) = B_A(0) = 3.7 * 10^8$ CFU/g is used, antibiotic therapy is administered 2 h post infection, and parameter values are fixed as shown in Table 1. To conduct the comparison study, the first experiment (Fig. 9a), is same as Fig. 8, i.e., only one dose of phages ($P = 7.4 * 10^8$ PFU/g) is administered 2 h post infection and it did not lead to a successful treatment. In the second experiment (Fig. 9b), we administer two doses of phages. The first dose ($P = 7.4 * 10^8$ PFU/g) is administered 2 h post infection. Then 10 h after the first dose, the second dose ($P = 1.8 * 10^{12}$ PFU/g, a relatively low value compared to Fig. 8b) is given. We found that even though the B_P density decreases quickly after the second dose of phage, it eventually rebound. This indicates we still fail the treatment. In the third experiment (Fig. 9c), we have three doses of phage therapy. The first two doses are administered as in the second experiment, i.e., the first dose ($P = 7.4 * 10^8$ PFU/g) is administered 2 h post infection. Then 10 h after the first dose, the second dose ($P = 1.8 * 10^{12}$ PFU/g, a relatively low value compared to Fig. 8b) is given. Now, 10 h after the second dose, we try the third dose ($P = 4.5 * 10^{11} < 1.8 * 10^{12}$ PFU/g), and see that we can obtain a successful treatment. Hence, we believe that the number of doses and the size of the dose of phages have significant impacts on the clearance of the bacterial infection.

In the above simulations, we have explored possible timing and dose size combination that allow the bacteria infection to be cleared. In each case, the final dose is as low as possible to result in the resolution of the infection. However, there maybe better, more effective timing/dose combinations that result in lower total phage dose. Future work will use optimal control to identify the best treatment plan for a given bacteria load.

6 Discussion

In this work, we have analyzed a prior model, developed in [50], for the use of combination antibiotic and phage therapy for the treatment of a systemic bacterial infection. We extended it by including immune response to circulating phages. While phages are not "infectious" to humans, they are a foreign substance in the body and will elicit an inflammatory response. Additionally, the innate immune response of the patient will clear some of the phages either through filtration or through phagocytosis. Therefore, we would like to see if this dynamic is important to consider for predicting the effectiveness of the combination therapy.

By utilizing equilibria analysis, we find that the model proposed by Rodriguez-Gonzalez, et al. [50] can have six possible steady-state cases for model outcomes. In addition, our analysis suggests that in some cases, the system might display bistable dynamics; i.e., the treatment outcomes can depend on initial conditions, determined by dose of drug or phage cocktail, or timing of any of these treatments, or the frequency of these treatments in combinations. Therefore, we numerically explores outcomes of treatment options using phage therapy in combination with antibiotic treatment in order to gain insights of how to optimize the treatment outcomes.

We performed a sensitivity analysis to determine which parameters are likely to affect the transient behavior and the overall outcome of the system. To that end, we found that two parameters were of the most interest. The one with the most biological meaning (ϕ) was investigated and found to have an effect on the outcome of the system. It will be important in future modeling work to better estimate the number of phage released during lysis while in a human host.

The timing of the phage treatment was also important for determining patient outcome. Because phages replicate inside the bacteria, the level of bacterial infection at the time of treatment initiation influences the effectiveness of the phage therapy. Repeated dosing with phages is also helpful in clearing the bacterial infection. Determining dosing protocols and quantifying the related risks will be important for future studies.

This initial investigation has been fruitful for understanding some of the competing dynamics observed in antibiotic/phage combination therapy, and has opened the work up to further questions and lines of research:

- Are there further interactions with the host immune system that need to be explored? (Innate/adaptive/filtering)
- Can we determine an optimal treatment strategy?
- Do different bacterial infections require different parameter values or are there other considerations that need to be made? Some bacteria have "broad spectrum" response to phages and some require treatment with very specific phages.
- How fast do bacteria develop or lose immunity to phages?
- What additional complications occur in immunocompromised individuals?

There is hope that phage therapy will usher in a new line of treatment for difficult bacterial infections but there are much work needed to understand the complex dynamics and to devise effective, broadly implementable treatment protocols.

Acknowledgments The work described herein was initiated during the Collaborative Workshop for Women in Mathematical Biology hosted by the Institute for Pure and Applied Mathematics (IPAM) at the University of California, Los Angeles in June 2019. Funding for the workshop was provided by IPAM, the Association for Women in Mathematics' NSF ADVANCE "Career Advancement for Women Through Research-Focused Networks" (NSF-HRD 1500481) and the Society for Industrial and Applied Mathematics. The authors thank the organizers of the IPAM-WBIO workshop (Rebecca Segal, Blerta Shtylla, and Suzanne Sindi) for facilitating this research. The authors would like to thank the two anonymous reviewers and the handling editor for their helpful comments and suggestions.

Appendix

Case IV. Phage-free equilibrium $(\mathcal{E}_{Pf}^+)$. Setting $P = 0$, at the steady-state we obtain the following equation system:

$$0 = r_p B_P \left(1 - \frac{B_{tot}}{K_c}\right)(1 - \mu_P) + \mu_A r_A B_A \left(1 - \frac{B_{tot}}{K_c}\right) - \frac{\epsilon I B_P}{1 + \frac{B_{tot}}{K_D}}, \qquad (12)$$

$$0 = r_A B_A \left(1 - \frac{B_{tot}}{K_c}\right)(1 - \mu_A) + \mu_P r_p B_P \left(1 - \frac{B_{tot}}{K_c}\right) - \frac{\epsilon I B_A}{1 + \frac{B_{tot}}{K_D}}$$

$$- K_{kill} \frac{A}{EC_{50} + A} B_A,$$

$$0 = \alpha I \left(1 - \frac{I}{K_I}\right) \left(\frac{B_{tot}}{B_{tot} + K_N}\right) + \beta I - dI.$$

By the last equation in (12), we have

$$\alpha \left(1 - \frac{I^+}{K_I}\right)\left(\frac{B_{tot}^+}{B_{tot}^+ + K_N}\right) = d - \beta. \tag{13}$$

Then rearranging (13), we obtain

$$B_{tot}^+ = \frac{(\beta - d)K_I}{\alpha I^+ K_I (d - \alpha - \beta)}. \tag{14}$$

Also by the first and second equations in (12), we obtain

$$\left(1 - \frac{B_{tot}^+}{K_c}\right) = \left(\frac{\epsilon I^+}{1 + \frac{B_{tot}^+}{K_D}}\right) \Big/ \left[r_p(1 - \mu_P) + \mu_A r_A \frac{B_A^+}{B_P^+}\right],$$

$$\left(1 - \frac{B_{tot}^+}{K_c}\right) = \left(\frac{\epsilon I^+}{1 + \frac{B_{tot}^+}{K_D}} + K_{kill}\frac{A}{EC_{50} + A}\right) \Big/ \left[r_A(1 - \mu_A) + \mu_P r_p \frac{B_P^+}{B_A^+}\right]. \tag{15}$$

By the equality of equations in (15), we have

$$\frac{\left(\dfrac{\epsilon I^+}{1 + \dfrac{B_{tot}^+}{K_D}}\right)}{\left[r_p(1 - \mu_P) + \mu_A r_A \dfrac{B_A^+}{B_P^+}\right]} = \frac{\left(\dfrac{\epsilon I^+}{1 + \dfrac{B_{tot}^+}{K_D}} + K_{kill}\dfrac{A}{EC_{50} + A}\right)}{\left[r_A(1 - \mu_A) + \mu_P r_p \dfrac{B_P^+}{B_A^+}\right]}. \tag{16}$$

Let $x = \dfrac{B_P^+}{B_A^+}$, $f(I^+) = \left(\dfrac{\epsilon I^+}{1 + \dfrac{B_{tot}^+}{K_D}}\right)$, $a_0 = r_p(1 - \mu_P)$, $b_0 = K_{kill}\dfrac{A}{EC_{50} + A}$, $c_0 = r_A(1 - \mu_A)$, $a_1 = \mu_A r_A$, and $c_1 = \mu_P r_p$. Then by (16), we obtain

$$\frac{f(I^+)}{\left[a_0 + a_1\dfrac{1}{x}\right]} = \frac{\left(f(I^+) + b_0\right)}{[c_0 + c_1 x]}. \tag{17}$$

Then rearranging it, we have

$$f(I^+)[c_0 + c_1 x] = \left(f(I^+) + b_0\right)\left[a_0 + a_1\frac{1}{x}\right]. \tag{18}$$

Multiplying both sides with x and rearranging we obtain

$$a_0 x^2 + a_1 x - a_2 = 0,$$

where

$$a_0 = c_1 f(I^+), a_1 = f(I^+)(c_0 - a_0) - b_0 a_0, a_2 = a_1(f(I^+) + b_0).$$

Therefore we get the steady-state ratio $x = \dfrac{B_P^+}{B_A^+}$ as follows:

$$\frac{B_P^+}{B_A^+} = \frac{-a_1 + \sqrt{a_1^2 + 4a_0 a_2}}{2a_0}, \tag{19}$$

where

$$a_0 = (\mu_P r_P) f(I^+), \tag{20}$$

$$a_1 = f(I^+)(r_A(1 - \mu_A) - r_p(1 - \mu_P)) - K_{kill} \frac{A}{EC_{50} + A} r_p(1 - \mu_P),$$

$$a_2 = a_1(f(I^+) + K_{kill} \frac{A}{EC_{50} + A}).$$

with $f(I) = \left(\dfrac{\epsilon I^+}{1 + \dfrac{B_{tot}^+}{K_D}} \right)$ and $B_{tot}^+ = \dfrac{(\beta - d)K_I}{\alpha I^+ K_I(d - \alpha - \beta)}.$

Also note that $B_P^+ = B_{tot}^+ - B_A^+$. Then by (19), we obtain

$$B_A^+ = \left(\frac{(\beta - d)K_I}{\alpha I^+ K_I(d - \alpha - \beta)} \right) \Big/ \left(\frac{-a_1 + \sqrt{a_1^2 + 4a_0 a_2}}{2a_0} \right), \tag{21}$$

where the expressions of a_i for $i = 0, 1, 2$ are given in (20). Therefore the system has at most one positive phage-free equilibrium $\mathcal{E}_{Pf}^+$.

Case V. Phage & Antibiotic Sensitive Bacteria (ASB)-free equilibrium $\mathcal{E}_{P\&ASBf} = (B_P^+, 0, A^+, 0, I^+)$. Setting $P = B_A = 0$, we obtain the following equation system:

$$\dot{B}_P = r_P B_P \left(1 - \frac{B_P}{K_c} \right)(1 - \mu_P) - \frac{\epsilon I B_P}{1 + \frac{B_P}{K_D}},$$

$$\dot{I} = \alpha I \left(1 - \frac{I}{K_I}\right)\left(\frac{B_P}{B_P + K_N}\right) - dI.$$

At the steady state, by the second equation, we have

$$\alpha \left(1 - \frac{I^+}{K_I}\right)\left(\frac{B_P^+}{B_P^+ + K_N}\right) = d. \qquad (22)$$

Rearranging it, we obtain

$$B_P^+ = \frac{K_N w}{1 - w}, \ with \ w = \frac{d}{\left(1 - \frac{I^+}{K_I}\right)}. \qquad (23)$$

By the first equation, we also have

$$B_{P1,2}^+ = \frac{(1 - \mu_p)r_p\left(-\frac{I^+}{K_c} + \frac{1}{K_D}\right)}{\frac{2(1 - \mu_p)r_p}{K_c K_D}}$$

$$\pm \frac{\sqrt{((1 - \mu_p)r_p)^2\left(-\frac{I^+}{K_c} + \frac{1}{K_D}\right)^2 - 4\frac{((1 - \mu_p)r_p)I^+((1 - \mu_p)r_p - \epsilon)}{K_c K_D}}}{\frac{2(1 - \mu_p)r_p}{K_c K_D}}.$$

$$(24)$$

Note that the equalities (23) and (24) are functions of I^+, and intersection of both equations give the equilibrium I^+ component of the equilibria of the system, and the other component of the equilibria can be found by substituting the component I^+ into the equation (23). It is clear that the system can have more than one phage & antibiotic sensitive bacteria-free equilibrium $\mathcal{E}_{P\&ASBf} = (B_P^+, 0, A^+, 0, I^+)$.

References

1. Abedon, S.: Kinetics of phage-mediated biocontrol of bacteria. Foodborne Pathog. Dis. **6**(7), 807–815 (2009)
2. Ankomah, P., Levin, B.R.: Exploring the collaboration between antibiotics and the immune response in the treatment of acute, self-limiting infections. Proceedings of the National Academy of Sciences **111**(23), 8331–8338 (2014)
3. Antibiotic/antimicrobial resistance: Center for disease control and prevention (September 10, 2018)

4. Austin, D., Anderson, R.: Studies of antibiotic resistance within the patient, hospitals and the community using simple mathematical models. Philos Trans R Soc Lond B Biol Sci **354**(1384), 721–738 (1999)
5. Austin, D.J., Kakehashi, M., Anderson, R.M.: The transmission dynamics of antibiotic-resistant bacteria: the relationship between resistance in commensal organisms and antibiotic consumption. Philos. Trans. R. Soc. Lond. B: Biol. Sci. **264**, 1629–1638 (1997)
6. Bergstrom, C.T., Lipsitch, M., Levin, B.R.: Natural selection, infectious transfer and the existence conditions for bacterial plasmids. Genetics **155**(4), 1505–1519 (2000)
7. Bergstrom, C.T., Lo, M., Lipsitch, M.: The transmission dynamics of antibiotic-resistant bacteria: the relationship between resistance in commensal organisms and antibiotic consumption. Proc. Natl. Acad. Sci. USA **101**, 13,285–13,290 (2004)
8. Bonhoeffer, S., Liptsitch, M., Levin, B.R.: Evaluating treatment protocols to prevent antibiotic resistance. Proc. Natl. Acad. Sci. USA **101**, 12,106–12,111 (1996)
9. Bootsma, M.C., Diekmann, J.O., Bonten, M.J.M.: Controlling methicillin-resistant Staphylococcus aureus: quantifying the effects of interventions and rapid diagnostic testing. Proc. Natl. Acad. Sci. USA **103**, 5620–5625 (2006)
10. Browne, C., Wang, M., Webb, G.F.: A stochastic model of nosocomial epidemics in hospital intensive care units. Electron. J. Qual. Theory Differ. Equ. **6**, 1–12 (2017)
11. Browne, C., Webb, G.F.: A nosocomial epidemic model with infection of patients due to contaminated rooms. Math Biosci Eng. **12**(4), 761–787 (2015)
12. Brun, R., Reichert, P., Künsch, H.R.: Practical identifiability analysis of large environmental simulation models. Water Resources Research **37**(4), 1015–1030 (2001)
13. Campbell, A.: Conditions for the existence of bacteriophage. Evolution **15**(2), 153–165 (1961)
14. Chamchod, F., Ruan, S.: Modeling methicillin-resistant Staphylococcus aureus in hospitals: transmission dynamics, antibiotic usage and its history. Theor. Biol. Med. Model. **9**, 25 (2012)
15. Chan B.k.and Sistrom, M., Wertz, J., Kortright, K., Narayan, D., Turner, P.: Phage selection restores antibiotic sensitivity in mdr Pseudomonas aeruginosa. Scientific Reports **6**(26717) (2016)
16. Chao, L., Levin, B., Stewart, F.: A complex community in a simple habitat: an experimental study with bacteria and phage. Ecology **58**(2), 369–378 (1977)
17. Clark, J.: Bacteriophage therapy: history and future prospects. Future Virol. **10**(4), 449–461 (2015)
18. Cooper, B., Medley, G., Stone, S., Kibbler, C., Cookson, B., Roberts, J., Duckworth, G., Lai, R., Ebrahim, S.: methicillin-resistant Staphylococcus aureus in hospitals and the community: stealth dynamics and control catastrophes. Proceedings of the National Academy of Sciences of the United States of America **101**(27), 10,223–10,228 (2004)
19. D'Agata, E.M., Horn, M.A., Ruan, S., Webb, G.F., Wares, J.R.: Efficacy of infection control interventions in reducing the spread of multidrug-resistant organisms in the hospital setting. PLoS One **7**(2), e30,170 (2012)
20. D'Agata, E.M., Magal, P., Olivier, D., Ruan, S., Webb, G.F.: Modeling antibiotic resistance in hospitals: the impact of minimizing treatment duration. Journal of Theoretical Biology **249**(3), 487–499 (2007)
21. D'Agata, E.M.C., Webb, G.F., Horn, M.A.: A mathematical model quantifying the impact of antibiotic exposure and other interventions on the endemic prevalence of vancomycin-resistant Enterococci. J. Infect. Disease. **192**, 2004–2011 (2005)
22. Davies, O.L.: Who publishes list of bacteria for which new antibiotics are urgently needed (2017). URL https://www.who.int/news-room/detail/27-02-2017-who-publishes-list-of-bacteria-for-which-new-antibiotics-are-urgently-needed
23. Dennehy, J.: What can phages tell us about host-pathogen coevolution? Int. J. Evol. Biol. **2012**(396165), 12 (2012)
24. Drusano, G.L., Vanscoy, B., Liu, W., Fikes, S., Brown, D., Louie, A.: Saturability of granulocyte kill of pseudomonas aeruginosa in a murine model of pneumonia. Antimicrobial agents and chemotherapy **55**, 2693–2695 (2011)

25. F. Jafri, H., Bonten, M.: Alternatives to antibiotics. Intensive Care Medicine **42**(12), 2034–2036 (2016)
26. Guenther, S., Huwyler, D., Richard, S., Loessner, M.: Virulent bacteriophage for efficient biocontrol of listeria monocytogenes in ready-to-eat foods. Appl. Environ. Microbiol. **75**(1), 93–100 (2009)
27. Hagens, S., Loessner, M.: Application of bacteriophages for detection and control of foodborne pathogens. Appl. Microbiol. Biotechnol. **76**(3), 513–519 (2007)
28. Hall, I.M., Barrass, I., Leach, S., Pittet, D., Hugonnet, S.: Transmission dynamics of methicillin-resistant Staphylococcus aureus in a medical intensive care unit. Journal of the Royal Society, Interface **9**(75), 2639–2652 (2012)
29. Hodyra-Stefaniak, K., Miernikiewicz, P., Drapała, J., Drab, M., Jończyk-Matysiak, E., Lecion, D., Kaźmierczak, Z., Beta, W., Majewska, J., Harhala, M., Bubak, B., Kłopot, A., Górski, A., Dąbrowska, K.: Mammalian host-versus-phage immune response determines phage fate in vivo. Scientific Reports **5**, 14,802 (2015)
30. Huang, Q., Horn, M.A., Ruan, S.: Modeling the effect of antibiotic exposure on the transmission of methicillin-resistant staphylococcus aureus in hospitals with environmental contamination. Mathematical Biosciences and Engineering **16**(5), 3641–3673 (2019)
31. Huang, Q., Huo, X., Miller, D., Ruan, S.: Modeling the seasonality of methicillin-resistant Staphylococcus aureus infections in hospitals with environmental contamination. J Biol Dyn. **13**(sup1), 99–122 (2018)
32. Huang, Q., Huo, X., Ruan, S.: Optimal control of environmental cleaning and antibiotic prescription in an epidemiological model of methicillin-resistant Staphylococcus aureus infections in hospitals. Math. Biosci. **311**, 13–30 (2019)
33. Kingwell, K.: Bacteriophage therapies re-enter clinical trials. Nature Reviews Drug Discovery **14**(8), 515–516 (2015)
34. Leung, C.Y.J., Weitz, J.S.: Modeling the synergistic elimination of bacteria by phage and the innate immune system. Journal of theoretical biology **429**, 241–252 (2017)
35. Levin, B., Stewart, F., Chao, L.: Resource-limited growth, competition, and predation: a model and experimental studies with bacteria and bacteriophage. Am. Nat. **111**(977), 3–24 (1977)
36. Levin, B.R., Bergstrom, C.T.: Bacteria are different: observations, interpretations, speculations, and opinions about the mechanisms of adaptive evolution in prokaryotes. Proceedings of the National Academy of Sciences **97**(13), 6981–6985 (2000)
37. Levin, B.R., Bull, J.J.: Population and evolutionary dynamics of phage therapy. Nature Reviews Microbiology **2**(2), 166–173 (2004)
38. Levin, B.R., Stewart, F.M.: The population biology of bacterial plasmids: a priori conditions for the existence of mobilizable nonconjugative factors. Genetics **94**(2), 425–443 (1980)
39. Lin, D.M., Koskella, B., Lin, H.C.: Phage therapy: An alternative to antibiotics in the age of multi-drug resistance. World journal of gastrointestinal pharmacology and therapeutics **8**, 162 (2017)
40. Lipsitch, M., Bergstrom, C.T., Levin, B.R.: The epidemiology of antibiotic resistance in hospitals: paradoxes and prescriptions. Proc. Natl. Acad. Sci. USA **97**, 1938–1943 (2000)
41. Luepke, K.H., Suda, K.J., Boucher, H., Russo, R.L., Bonney, M.W., Hunt, T.D., Mohr III, J.F.: Past, present, and future of antibacterial economics: increasing bacterial resistance, limited antibiotic pipeline, and societal implications. Pharmacotherapy: The Journal of Human Pharmacology and Drug Therapy **37**, 71–84 (2017)
42. Luria, S.E., Delbrück, M.: Mutations of bacteria from virus sensitivity to virus resistance. Genetics **28**, 491 (1943)
43. Meyer, J., Dobias, D., Weitz, J., Barrick, J., Quick, R., Lenski, R.: Repeatability and contingency in the evolution of a key innovation in phage lambda. Science **335**(6067), 428–432 (2012)
44. Pincus, N., Reckhow, J., Saleem, D., Jammeh, M., Datta, S., Myles, I.: Strain specific phage treatment for staphylococcus aureus infection is influenced by host immunity and site of infection. PLoS ONE **10**(4), 1–16 (2015)

45. Potera, C.: Phage renaissance: new hope against antibiotic resistance. Environ Health Perspect. **121**, a48–53 (2013)
46. Rea, K., Dinan, T.G., Cryan, J.F.: The microbiome: A key regulator of stress and neuroinflammation. Neurobiol Stress **4**, 23–33 (2016)
47. Reardon, S.: Phage therapy gets revitalized. Nature **510**(7503), 15–16 (2014)
48. Reutershan, J., Basit, A., Galkina, E.V., Ley, K.: Sequential recruitment of neutrophils into lung and bronchoalveolar lavage fluid in LPS-induced acute lung injury. American Journal of Physiology-Lung Cellular and Molecular Physiology **289**, L807–L815 (2005)
49. Roach, D.R., Leung, C.Y., Henry, M., Morello, E., Singh, D., Di Santo, J.P., Weitz, J.S., Debarbieux, L.: Synergy between the host immune system and bacteriophage is essential for successful phage therapy against an acute respiratory pathogen. Cell host & microbe **22**(1), 38–47 (2017)
50. Rodriguez-Gonzalez, R.A., Leung, C.Y., Chan, B.K., Turner, P.E., Weitz, J.S.: Quantitative models of phage-antibiotic combination therapy. Msystems **5**(1) (2020)
51. Seo, J., Seo, D., Oh, H., Jeon, S., Oh, M.H., Choi, C.: Inhibiting the growth of escherichia coli O157:H7 in beef, pork, and chicken meat using a bacteriophage. Korean J. Food Sci. Anim. Resour. **2**(2), 186–93 (2016)
52. Stewart F. M., Levin, B.R.: The population biology of bacterial plasmids: a priori conditions for the existence of conjugationally transmitted factors. Genetics **87**(2), 209–228 (1977)
53. Thiel, K.: Old dogma, new tricks–21st century phage therapy. Nat. Biotechnol. **22**(1), 31–36 (2004)
54. Tiwari, B., Kim, S., Rahman, M., Kim, J.: Antibacterial efficacy of lytic pseu- domonas bacteriophage in normal and neutropenic mice models. J. Microbiol. **49**(6), 994–999 (2011)
55. Torres, M., Wang, J., Yannie, P., Ghosh, S., Segal, R., Reynolds, A.: Identifying important parameters in the inflammatory process with a mathematical model of immune cell influx and macrophage polarization. PLoS Comput Biol **15**(7), 497–503 (2019)
56. Torres-Barceló, C., Franzon, B., Vasse, M., Hochberg, M.E.: Long-term effects of single and combined introductions of antibiotics and bacteriophages on populations of pseudomonas aeruginosa. Evolutionary applications **9**, 583–595 (2011)
57. Trigo, G., Martins, T., Fraga, A., Longatto-Filho, A., Castro, A., Azeredo, J., Pedrosa, J.: Phage therapy is effective against infection by mycobacterium ulcerans in a murine footpad model. PLoS Negl. Trop. Dis **7**(4), e2183 (2013)
58. Wang, J., Wang, L., Magal, P., Wang, Y., Zhuo, S., Lu, X., Ruan, S.: Modelling the transmission dynamics of Meticillin-resistant Staphylococcus aureus in Beijing Tongren Hospital. Journal of Hospital Infection **79**(4), 302–308 (2011)
59. Wang, L., Ruan, S.: Modeling nosocomial infections of methicillin-resistant Staphylococcus aureus with environment contamination. Scientific Reports **7** (2017)
60. Wang, X., Xiao, Y., Wang, J., Lu, X.: A mathematical model of effects of environmental contamination and presence of volunteers on hospital infections in China. Journal of Theoretical Biology **293**, 161–173 (2012)
61. Wang, X., Xiao, Y., Wang, J., Lu, X.: Stochastic disease dynamics of a hospital infection model. Mathematical Biosciences **241**(1), 115–124 (2013)
62. Webb, G.F.: Individual based models and differential equations models of nosocomial epidemics in hospital intensive care units. Discrete & Continuous Dynamical Systems-Series B **22**(3) (2017)
63. Webb, G.F., D'Agata, E.M.C., Magal, P., Ruan, S.: A model of antibiotic resistant bacterial epidemics in hospitals. Proc. Natl. Acad. Sci. USA **102**, 13,343–13,348 (2005)
64. Weitz, J., Hartman, H., Levin, S.: Coevolutionary arms races between bacteria and bacteriophage. Proc. Natl. Acad. Sci. USA. **102**(27), 9535–9540 (2005)
65. World Health Organization: Antimicrobial resistance – global action plan (2015)
66. Zhang, Z., Louboutin, J.P., Weiner, D.J., Goldberg, J.B., Wilson, J.M.: Human airway epithelial cells sense pseudomonas aeruginosa infection via recognition of flagellin by toll-like receptor 5. Infection and immunity **73**, 7151–7160 (2005)

Mathematical Modeling of Retinal Degeneration: Aerobic Glycolysis in a Single Cone

Erika Tatiana Camacho, Atanaska Dobreva, Kamila Larripa, Anca Rădulescu, Deena Schmidt, and Imelda Trejo

Abstract Cell degeneration, including that resulting in retinal diseases, is linked to metabolic issues. In the retina, photoreceptor degeneration can result from imbalance in lactate production and consumption as well as disturbances to pyruvate and glucose levels. To identify the key mechanisms in metabolism that may be culprits of this degeneration, we use a nonlinear system of differential equations to mathematically model the metabolic pathway of aerobic glycolysis in a single cone photoreceptor. This model allows us to analyze the levels of lactate, glucose, and pyruvate within a single cone cell. We perform numerical simulations, use available metabolic data to estimate parameters and fit the model to this data, and conduct a sensitivity analysis using two different methods (LHS/PRCC and eFAST) to identify pathways that have the largest impact on the system. Using bifurcation techniques,

E. T. Camacho (✉)
School of Mathematical and Natural Sciences, Arizona State University, Glendale, AZ, USA
e-mail: Erika.Camacho@asu.edu

A. Dobreva
School of Mathematical and Natural Sciences, Arizona State University, Glendale, AZ, USA

Department of Mathematics, North Carolina State University, Raleigh, NC, USA
e-mail: adobreva@asu.edu

K. Larripa
Department of Mathematics, Humboldt State University, Arcata, CA, USA
e-mail: kamila.larripa@humboldt.edu

A. Rădulescu
Department of Mathematics, State University of New York at New Paltz, New Paltz, NY, USA
e-mail: radulesa@newpaltz.edu

D. Schmidt
Department of Mathematics and Statistics, University of Nevada Reno, Reno, NY, USA
e-mail: drschmidt@unr.edu

I. Trejo
Theoretical Biology and Biophysics, Los Alamos National Laboratory, Los Alamos, NM, USA
e-mail: imelda@lanl.gov

© The Association for Women in Mathematics and the Author(s) 2021   135
R. Segal et al. (eds.), *Using Mathematics to Understand Biological Complexity*,
Association for Women in Mathematics Series 22,
https://doi.org/10.1007/978-3-030-57129-0_7

we find that the system has a bistable regime, biologically corresponding to a healthy versus a pathological state. The system exhibits a saddle node bifurcation and hysteresis. This work confirms the necessity for the external glucose concentration to sustain the cell even at low initial internal glucose levels. It also validates the role of β-oxidation of fatty acids which fuel oxidative phosphorylation under glucose- and lactate-depleted conditions, by showing that the rate of β-oxidation of ingested outer segment fatty acids in a healthy cone cell must be low. Model simulations reveal the modulating effect of external lactate in bringing the system to steady state; the bigger the difference between external lactate and initial internal lactate concentrations, the longer the system takes to achieve steady state. Parameter estimation for metabolic data demonstrates the importance of rerouting glucose and other intermediate metabolites to produce glycerol 3-phosphate (G3P), thus increasing lipid synthesis (a precursor to fatty acid production) to support their high growth rate. While a number of parameters are found to be significant by one or both of the methods for sensitivity analysis, the rate of β-oxidation of ingested outer segment fatty acids is shown to consistently play an important role in the concentration of glucose, G3P, and pyruvate, whereas the extracellular lactate level is shown to consistently play an important role in the concentration of lactate and acetyl coenzyme A. The ability of these mechanisms to affect key metabolites' variability and levels (as revealed in our analyses) signifies the importance of inter-dependent and inter-connected feedback processes modulated by and affecting both the RPE's and cone's metabolism.

Keywords Retina · Photoreceptors · Aerobic glycolysis · β-oxidation and differential equations

1   Introduction

Photoreceptors are the sensory cells of the eye, and they are the most energetically demanding cells in the body [52]. Photoreceptors have the most essential role in vision, absorbing light photons and processing them to electrical signals that can be transmitted to the brain. Therefore, vision deterioration or blindness occurs if the vitality and functionality of photoreceptors are compromised. In order to understand how to mitigate such pathological cases, it is essential to first obtain a firm grasp of processes that ensure the health of photoreceptors. The factor of upmost importance for photoreceptor vitality and functionality is metabolism.

To maintain their high metabolic demands and prevent accumulation of photo-oxidative product, the photoreceptors undergo constant renewal and periodic shedding of their fatty acid-rich outer segment (OS) discs. Aerobic glycolysis is integral to the renewal process. It facilitates the production of energy and the synthesis of phospholipids, both which are required for OS renewal. Phagocytosis of the shed OS by the retinal pigment epithelium (RPE) contributes to the creation of intermediate metabolites fundamental for photoreceptor energy production via β-oxidation [1].

Understanding the dynamics of glucose and lactate levels in aerobic glycolysis in a single cone cell is essential to maintain cone functionality and hence to preserve central vision. Studies in rod-less retinas have shown that maintaining functional cones even when 95% are gone may stop blindness [11, 30]. The purpose of this study is to analyze the key mechanisms affecting the levels of glucose, pyruvate, and lactate in a single cone cell via a first approximation mathematical model, with the goal of gaining insight into the interplay of glucose consumption and lactate production and consumption that may affect normal cone function.

1.1   Biological Background and Modeling Assumptions

1.1.1   Photoreceptors and Retinal Pigment Epithelium (RPE)

Photoreceptors are specialized neurons that convert light into electrical signals that can be interpreted by the brain [37]. There are two types of photoreceptor cells: rods and cones. Cones are densely packed in the center of the retina and are responsible for color vision and high acuity. Rods have high sensitivity to light, are distributed on the outer edges of the retina, and are responsible for night and peripheral vision. In the human retina, there are approximately 90 million rod cells and 4.5 million cone cells [15], making rods twenty times more prevalent than cones. In the mature human retina (by about age 5 or 6), there are no spontaneous births of photoreceptors, making their preservation and vitality critical [10]. Photoreceptor shedding and renewal of their OS has been considered as a type of death and birth process, as it is the mechanism by which photoreceptors discard unwanted elements (e.g., accumulated debris or toxic photo-oxidative compound in shed OS discs) and renew themselves through the recycling of various products. This process is a measurement of the photoreceptor's energy uptake and consumption and associated metabolism [11, 12]. The shedding and renewal process and the associated metabolism of photoreceptors involve the RPE. The photoreceptors and the RPE work as a functional unit; glucose is transported from the RPE to the photoreceptors for their metabolism, lactate produced by photoreceptors and other retina cells is shuttled to the RPE for its metabolism, and the RPE mediates the phagocytosis of photoreceptor OS and recycling of fatty acids from these OS discs which are utilized in oxidative phosphorylation (OXPHOS) in the production of acetyl coenzyme A (ACoA); see Fig. 1. However, as a first approximation we will not consider the role of the RPE but instead integrate the feedback mechanisms back into the cone cell via β-oxidation and external lactate transport.

The RPE lies between the choroid and a layer of photoreceptors. In addition to functioning as the outer blood retinal barrier and transporting glucose to photoreceptor cells through GLUT1 (a facilitated glucose transporter), the RPE is involved in the phagocytosis of photoreceptor OS discs [57]. It serves as the principal pathway for the exchange of metabolites and ions between the choroidal blood supply and the retina [14]. Müeller cells are a layer of retinal glial cells and also

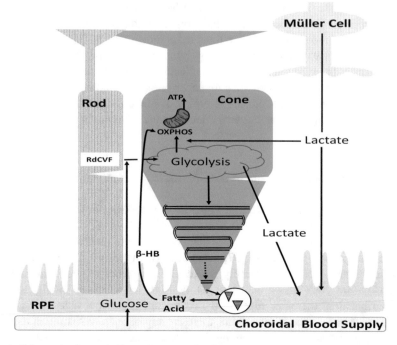

Fig. 1 Schematic of metabolic pathways and substrate sources in the photoreceptors. This schematic shows that glucose and lactate flow between the cone photoreceptor and the RPE cell layer. It illustrates the contribution of glycolysis in providing energy to the cone cell and its role in helping generate cone outer segments. β-Hydroxybutyrate oxidation (β-HB) comes from oxidation of fatty acids from the shed outer segments, so that under starvation or low glucose levels they can be used as oxidative substrates [1]

provide support to photoreceptors. They can release lactate which is metabolized by photoreceptors [41] and store glycogen which can be broken down to glucose. A thorough investigation should consider the interaction of the three cell types. However, in this work we consider, as a first step, a single cone photoreceptor in the human retina and model the metabolic pathways present. This analysis provides the foundation for a future application of the model: prediction of the interplay of metabolites from three cell types (RPE, photoreceptors, and Müeller) coexisting in the retina.

1.1.2 Glycolysis and Oxidatative Phosphorylation

Photoreceptors are responsible for the majority of the energy consumption in the retina [38, 50]. Active transport of ions against their electrical and concentration gradients in neurons is required to repolarize the plasma membrane after depolarization, and this process is what consumes the most energy in photoreceptors

[37, 52]. Moreover, the continual renewal and periodic shedding of OS [56] is also an extremely energetically demanding process.

All life on Earth relies on adenosine triphosphate (ATP) in energy transfer. ATP is produced via two pathways, oxidative phospholylation and glycolysis. Glycolysis, through a series of reactions (described in detail in Sect. 2.1), converts one molecule of glucose into two molecules of pyruvate, yielding two net molecules of ATP. If oxygen is present, pyruvate is typically converted to ACoA and enters the tricarboxylic acid (TCA) cycle, generating 32 net ATP molecules through OXPHOS. If oxygen is scarce, or if a cell has been metabolically reprogrammed, pyruvate is instead converted to lactate. However, photoreceptor cells use both pathways for energy production in the presence of oxygen with the vast majority of pyruvate being converted into lactate. In other words, despite only producing two molecules of ATP (versus 32 via OXPHOS), photoreceptors go through glycolysis as well as OXPHOS.

Glucose serves as the primary fuel in photoreceptors [13] and is broken down through aerobic glycolysis (glycolysis even in the presence of oxygen), termed the Warburg effect [2]. The Warburg effect has long been noted as a hallmark of tumors [23], but is also present in healthy tissue, particularly if their biosynthetic demands are high. Aerobic glycolysis maintains high fluxes through anabolic pathways and creates excess carbon which can be exploited for generation of nucleotides, lipids, and proteins, or diverted to other pathways branching from glycolysis, such as the pentose phosphate pathway and Kennedy pathway [32].

During glycolysis, glucose is transported into the cell. Rod-derived cone viability factor (RdCVF), which is secreted by rod photoreceptors, accelerates the uptake of glucose by cones through its binding with the glucose transporter complex 1/Basigin-1 (GLUT1/BSG-1) and stimulates aerobic glycolysis [3]. RdCVF also protects cones from degeneration [28, 53]. When glucose is in short supply, photoreceptors have the ability to take up and metabolize lactate [41].

1.1.3 Lactate Secretion and Consumption

Photoreceptors can produce lactate from pyruvate and secrete it out of the cell or consume external lactate and convert it to pyruvate for OXPHOS if there is too much lactate in the extracellular space. The influx of lactate from the extracellular space would almost certainly slow the rate of glycolysis in the cell because any resulting higher intracellular lactate concentration shifts the lactate dehydrogenase (LDH)-catalyzed reaction equilibrium toward a higher $NADH/NAD^+$ ratio. Under normal conditions, retinal cells oxidize cytosolic NADH to NAD^+ (via the reduction/conversion of pyruvate to lactate in order to regenerate the NAD^+). This lactate is transported out of the cell, thus increasing the amount of extracellular lactate. When glucose is low, such as during hypoglycemia or aglycemia conditions or hypoxia, oxidation of external lactate and fatty acids (via β-oxidation) to generate ACoA and thus produce energy (ATP) is favored [51]. When the photoreceptor cell undergoes

OXPHOS, it makes citrate which provides an inhibitory feedback to glycolysis when other intermediates for ATP production are high, indicating additional glucose is not needed.

Glucose from the choroidal blood passes through the RPE to the retina where photoreceptors convert it to lactate, and in return, photoreceptors then export lactate as fuel for the RPE and for neighboring cells [26]. It has been hypothesized that photoreceptors also take up lactate for energy under low glucose levels. In humans, insufficient lactate transported out of the cone and rod cells for RPE consumption can suppress transport of glucose by the RPE. In such a case, the RPE takes glucose for its metabolism thereby decreasing the amount of glucose that is transported to the photoreceptors. Thus, lactate secretion for RPE consumption and external lactate consumption by photoreceptors is a balance process.

1.1.4 Modeling Assumptions

Our model consolidates some of the steps in the glycolytic pathway in a single cone, for simplicity. Glucose is initially transported into the cell, and the rate of transport is amplified by the release of RdCVF from rods. The rate of transport is gradient dependent and modulated by the difference in the amount of glucose inside and outside the cell. The next step is the conversion of glucose in the cell into glucose-6-phosphate (G6P) by the enzyme hexokinase 2. This phosphorylation also works to trap glucose in the cell's cytosol. Some G6P is diverted to the pentose phosphate pathway (not included in our model), while the rest moves through the glycolytic pathway. The enzyme phosphofructokinase (PFK) converts fructose-6-phosphate to fructose 1,6-biphosphate (not explicitly included in our model). This in turn is cleaved into two sugar molecules, one of which is dihydroxyacetone phosphate (DHAP), the substrate for the next reaction. DHAP is converted to glycerol-3-phosphate (G3P) in the Kennedy pathway and glyceraldehyde-3-phosphate (GAP) in the glycolytic pathway. The latter metabolite is not explicitly considered in our model. A number of sequential reactions occur, with the ultimate step aided by the enzyme pyruvate kinase, resulting in pyruvate. Since our model considers a single cone, we use the presence of the metabolite concentration [G3P] with an appropriate scaling factor as a proxy for the amount of RdCVF synthesized by the rods. RdCVF accelerates glucose uptake in cones [3, 28, 53].

Specifically, our model incorporates the uptake and consumption of glucose, the production of G3P and pyruvate, and key consecutive chemical reactions in the cone cell involving lactate, ACoA, and citrate. Pyruvate is converted to lactate, which is then transported out of the cell. A portion of pyruvate is also transferred to the mitochondria; there it is converted to ACoA and goes through OXPHOS, creating citrate, which leads to the production of ATP but also negatively regulates the glycotic pathway. Citrate inhibits phosphofructokinase (PFK) which slows down the production of G3P and pyruvate. G3P leads to the production of lipids which are used to create OS that are shed and phagocytized periodically. The fatty acids from

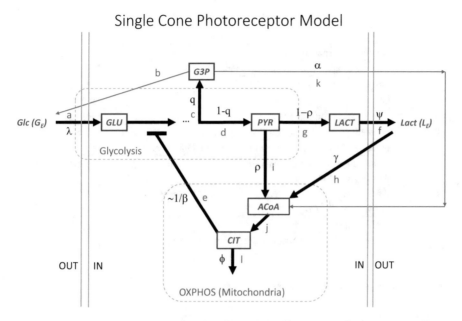

Fig. 2 Flow diagram of the key metabolic pathways within a single cone photoreceptor. Parameters corresponding to each pathway are labeled with black letters while metabolic pathways are labeled with blue letters (a-l). Parameters are described in Table 2. Brief descriptions of each pathway are given in Table 1 and are described in detail in Sect. 2.1

the shed OS can feed back into the cone cell as β-hydroxybutyrate (β-HB) which serves as a substrate for ACoA production. Through the process of β-oxidation fatty acids in the RPE result in β-HB. The specific pathways are outlined in Fig. 2, and specific evidence is presented for each pathway in detail below.

2   Mathematical Model

We model six key steps in the glycolytic pathway as a system of six nonlinear ordinary differential equations that describes metabolic pathways in a single cone. Specifically, we track the temporal dynamics of the following six concentrations in the cell: internal glucose ([G]), glycerol-3-phosphate ([G3P]), pyruvate ([PYR]), lactate ([LACT]), acetyl coenzyme A ([ACoA]), and citrate ([CIT]). The chemical reactions and up- and down-regulations included in this model are illustrated in Fig. 2 and listed in Table 1. In Sect. 2.1, we discuss the biological basis for each interaction pathway used in the model. In Sects. 2.2 and 2.3, we give the model equations and parameter values used, respectively.

Table 1 Description of metabolic pathways in the model

Pathway	Description	References
a	Gradient transport of glucose	[33, 40]
b	Glucose uptake; without and with RdCVF	[3, 10, 53]
c	Glycolytic flow diverted to G3P	[43, 46]
d	Glycolytic flow diverted to pyruvate	[6]
e	Glycolysis inhibition by citrate	[6]
f	Gradient gating mechanism to transport lactate out of the cell	[5, 8, 9, 22, 25]
g	Fraction of pyruvate concentration converted into lactate	[18, 41, 55]
h	Gradient gating mechanism to transport lactate into the cell for ACoA production	[6, 20]
i	Fraction of pyruvate concentration converted into ACoA	[36, 47]
j	Conversion of ACoA to citrate	[47]
k	β-HB utilized in production of ACoA	[1]
l	Diversion of citrate to the cytosol and other metabolic pathways	[47]

2.1 Kinetic Pathways in the Model

Here, we provide details of all model pathways shown in Fig. 2 and described in Table 1. These pathways represent a reduced system, with some pathways omitted and elements implicitly modeled via proxies. There are multiple intermediates produced in glycolysis and oxidative phosphorylation which are not explicitly considered in this work. In order to focus on production and consumption of glucose, lactate, and pyruvate in a single cone cell we reduce the system to its most essential components and pathways.

Pathway a: gradient transport of glucose

In the retina, sodium independent glucose transporters (GLUTs) transport glucose by facilitated diffusion down its concentration gradient [40]. GLUT1 is found in human photoreceptor outer segments [33]. We model this pathway by considering the difference between external glucose concentration (the parameter G_E in our model) and the internal glucose concentration, the variable [G]. The parameter λ is a constant of proportionality that governs the rate of glucose uptake based on the concentration gradient.

Pathway b: glucose uptake without and with stimulation of GLUT1 by RdCVF

Rod-derived cone viability factor (RdCVF) is secreted in a paracrine manner by rod photoreceptors and protects cones from degeneration [3, 53]. It binds with the GLUT1/BSG-1 complex to activate GLUT1 and accelerates the entry of glucose into the cone. We use [G3P] together with an appropriate scaling factor incorporated into δ as a proxy for the RdCVF that is synthesized by the rods. G3P is needed for phospholipid synthesis resulting in the renewal of photoreceptor OS [10]. Since for every cone cell there are approximately 20 rods in the human retina and G3P in our model is a measurement of the cone OS, we scale concentrations to account for rods' secretion of RdCVF that accelerates glucose uptake and supports cone vitality.

Note that the parameter n is included in this term so that even in the absence of RdCVF or our proxy for rods, glucose is passively transported down its gradient (bidirectionally) into the intracellular space of the cone cell. Thus, λn is the glucose uptake rate of our cone cell in the absence of RdCVF. The uptake of glucose due to RdCVF is an allosteric reaction, and therefore there is a binding time requirement for the enzyme to catalyze the reaction. We therefore use a Hill type function with a Hill coefficient of 2 to model the sigmoidal response [10, 42].

Pathway c: glycolytic flow diverted to G3P

The sequence of reactions leading to G3P are glucose to glucose-6 phosphate to fructose 6-phosphate to fructose 1,6-biphosphate to dihydroxyacetone phosphate (DHAP), and then G3P. Figure 2 indicates the many intermediate steps which are skipped with the ellipsis in the diagram. We model the conversion of glucose to G3P with a Hill type function where $V_{[G3P]}$ is the maximal rate of conversion (controlled by the rate-limiting allosteric enzyme PFK described above) of glucose to G3P and $K_{[G3P]}$ is the concentration of the ligand that gives half-maximal activity [46].

Pathway d: glycolytic flow diverted to pyruvate

We skip intermediate reactions to focus on the key metabolites of interest; glucose, pyruvate, and lactate. We take a similar approach as in pathway c and consider glucose to be the substrate in the reaction resulting in the production of pyruvate. We can infer from known aerobic glycolysis that the substrate (in this case glucose) which is not converted to G3P is converted to pyruvate [6]. Thus, a fraction q of glucose gets converted to G3P, while the remaining fraction $1 - q$ gets converted to pyruvate.

Pathway e: glycolysis inhibition by citrate

The flux through the glycolytic pathway must be responsive to conditions both inside and outside the cell, and the enzyme phosphofructokinase (PFK) is a key element in this control. PFK is inhibited by citrate, which enhances the allosteric inhibitory effect of ATP [6]. Elevated citrate levels indicate that biosynthetic precursors are readily available and additional glucose should not be degraded. The form of the function capturing this inhibition is reciprocal to the concentration of citrate and is multiplied to the metabolic reactions that involve glucose as a substrate down the glycolysis pathway (i.e., the reactions that produce G3P and pyruvate).

Pathway f: gradient gating mechanism to transport lactate out of the cell

Monocarboxylate transport proteins (MCT) are a family of plasma membrane transporters and allow lactate, pyruvate, and ketone bodies to be actively transported across cell membranes [5]. The RPE expresses various isoforms of the MCT transporter [9], as do the photoreceptors, Müeller cells, and the inner blood-retinal barrier. Inhibition of MCT results in retinal function loss [9], mainly due to lactate accumulation in the extracellular space. The lactate transport rate is dependent on pH, temperature, and concentration of internal lactate relative to external cellular lactate [25]. MCTs faciliate lactate transport down the concentration and pH gradients [8]. MCT1 is particularly important for reducing conditions of intracellular acidification when glycotic flux is high [22]. MCT1 transports lactate out of the photoreceptors and into the RPE. We incorporate this process in our model with pathway **f**.

Lactate flux out of the cell depends on the concentration gradient, pH, and temperature. We model this using a gating function $f([LACT])$; see Equation 8. For large binding affinity of lactate transporter (i.e., for large k values), if the external lactate concentration L_E exceeds the internal lactate concentration [LACT], the gate closes, and the external lactate is directed to OXPHOS via function $h([LACT])$ to produce ACoA; see Equation 9. The height of this function represents the maximal flux possible, physiologically limited by the concentration and expression of MCT1.

Pathway g: fraction of pyruvate concentration converted into lactate

Pyruvate is converted to lactate in glycolysis. This metabolic reaction is promoted by increased expression of the enzyme lactate dehydrogenase A (LDHA) and inactivation of pyruvate dehydrogenase [18, 41]. The conversion and direction of the reaction from pyruvate to lactate depends on lactate dehydrogenase subtypes; photoreceptors express LDHA which favors the production of lactate from pyruvate [10].

Pathway h: gradient gating mechanism to transport lactate into the cell for ACoA production

Pyruvate dehydrogenase complex (PDC) converts pyruvate to ACoA. The consumption of lactate back into the cell depends on a gating mechanism modulated by the pH levels and the lactate gradient inside and outside the cell. While LDHA converts pyruvate to lactate, lactate dehydrogenase B (LDHB) converts lactate to pyruvate. The latter reaction involves external lactate and the newly acquired pyruvate does not convert back to lactate but rather goes into the mitochondria where it becomes a substrate in the production of ACoA. The conversion of lactate to pyruvate and vice versa also depends on NAD^+ and NADH levels as they can drive things in one direction or another. When lactate is used as an energy source, lactate carbon is ultimately inserted into the TCA cycle in the mitochondria.

Glycolysis and gluconeogenesis are coordinated so that within one cell, one pathway is relatively inactive while the other is highly active. The rate of glycolysis is governed by the concentration of glucose whereas the rate of gluconeogenesis is governed by the concentration of lactate [20]. Inhibition of the enzyme PFK (which drives glycolysis) and abundance in [CIT] activates gluconeogenesis [6]. Rather than modeling all steps of gluconeogenesis, we let external lactate feed directly to ACoA and do not track its passage through pyruvate. This mechanism consolidates entry of lactate into the cell.

Pathway i: fraction of pyruvate converted into ACoA

After pyruvate is produced, its flux branches off and a fraction ρ of pyruvate is transferred to the mitochondria by the mitochondrial pyruvate carrier and converted into ACoA. During glycolysis, the mitochondrial pyruvate dehydrogenase complex catalyzes the oxidative decarboxylation of pyruvate to produce ACoA [36, 47].

Pathway j: conversion of ACoA to citrate

In the mitochondria, the enzyme citrate synthase catalyzes the conversion of ACoA and oxaloacetate into citrate [47].

Pathway k: fatty acids utilized in production of ACoA

During the shedding and subsequent phagocytosis of the OS, a source of fatty acids is created [1]. This itself can be used for metabolism, and feeds directly into ACoA. As depicted in pathway **k**, we use G3P as a proxy for the substrates that are created through β-oxidation (the process by which fatty acid molecules are broken down in the mitochondria to generate ACoA). G3P is converted to lipids which form the photoreceptor's OS that eventually get shed and become phagolysosomes containing fatty acids. These fatty acids can be oxidized and generate ACoA. ACoA leads to the production of β-Hydroxybutyrate (β-HB) via ketogenesis which can be used as an oxidative substrate in the TCA cycle when glucose is low. In conditions of glucose starvation, fatty acids are released, broken down, oxidized, and used to produce ketones that can be used to fuel the cone cell. In our mathematical model, we do not directly model ketogenesis but instead G3P serves as a proxy for β-oxidation of fatty acids from ingested OS.

Pathway l: diversion of citrate to the cytosol and other metabolic pathways

Citrate in the mitochondria can be oxidized via the TCA cycle, or it can be moved to the cytosol to be cleaved by ATP citrate lyase, which regenerates ACoA and oxaloacetate. This pathway redirects ACoA away from the mitchondria under conditions of glucose excess [47]. It reduces the glycolytic flux coming into the TCA cycle and signals the cone cell that ATP is high and there is no need for glucose metabolism.

2.2 Model Equations

Following the flow diagram given in Fig. 2, we apply mass-action Michaelis-Menten kinetics and allosteric regulations to the relevant parts of the variable interactions to yield the resulting system of equations:

$$
\frac{d[G]}{dt} = \overbrace{\lambda(G_E - [G])}^{a} \overbrace{\left(\frac{V_{[G]}(\delta[G3P])^2}{K_{[G]}^2 + (\delta[G3P])^2} + n \right)}^{b}
$$

$$
- \left(\overbrace{\frac{q V_{[G3P]}[G]^2}{K_{[G3P]}^2 + [G]^2}}^{c} + \overbrace{\frac{(1-q)V_{[PYR]}[G]^2}{K_{[PYR]}^2 + [G]^2}}^{d} \right) \overbrace{\left(\frac{1}{1 + \beta[CIT]} \right)}^{e} \tag{1}
$$

$$
\frac{d[G3P]}{dt} = \overbrace{\frac{q V_{[G3P]}[G]^2}{K_{[G3P]}^2 + [G]^2}}^{c} \overbrace{\left(\frac{1}{1 + \beta[CIT]} \right)}^{e} - \overbrace{\alpha[G3P]}^{k} \tag{2}
$$

$$\frac{d[\text{PYR}]}{dt} = \overbrace{\frac{(1-q)V_{[\text{PYR}]}[G]^2}{K_{[\text{PYR}]}^2 + [G]^2}}^{d} \overbrace{\left(\frac{1}{1+\beta[\text{CIT}]}\right)}^{e}$$

$$- \overbrace{\frac{(1-\rho)V_{[\text{LACT}]}[\text{PYR}]}{K_{[\text{LACT}]} + [\text{PYR}]}}^{g} - \overbrace{\frac{\rho V_{[\text{ACoA}]}[\text{PYR}]}{K_{[\text{ACoA}]} + [\text{PYR}]}}^{i} \qquad (3)$$

$$\frac{d[\text{LACT}]}{dt} = \overbrace{\frac{(1-\rho)V_{[\text{LACT}]}[\text{PYR}]}{K_{[\text{LACT}]} + [\text{PYR}]}}^{g} - \overbrace{f([\text{LACT}])([\text{LACT}] - L_E)}^{f} \qquad (4)$$

$$\frac{d[\text{ACoA}]}{dt} = \overbrace{\frac{\rho V_{[\text{ACoA}]}[\text{PYR}]}{K_{[\text{ACoA}]} + [\text{PYR}]}}^{i} + \overbrace{h([\text{LACT}])(L_E - [\text{LACT}])}^{h}$$

$$- \overbrace{\frac{V_{[\text{CIT}]}[\text{ACoA}]}{K_{[\text{CIT}]} + [\text{ACoA}]}}^{j} + \overbrace{\alpha[\text{G3P}]}^{k} \qquad (5)$$

$$\frac{d[\text{CIT}]}{dt} = \overbrace{\frac{V_{[\text{CIT}]}[\text{ACoA}]}{K_{[\text{CIT}]} + [\text{ACoA}]}}^{j} - \overbrace{\phi[\text{CIT}]}^{l} \qquad (6)$$

The model consists of 25 parameters defining various metabolic kinetic processes affecting internal [G], [PYR], and internal [LACT] within a cone cell; see Table 2. Since we are not incorporating the RPE and the rod cells, we consider three intermediate metabolites, G3P, ACoA, and citrate, that affect energy production and are sources of feedback mechanisms. The former two provide feedback mechanisms for glucose and fatty acids (in the form of β-HB) to enter the cone cell. They are proxies for mechanisms being mediated by the RPE and rod cells. The metabolite G3P in a healthy cone cell can be used to approximate the rods that synthesize RdCVF as well as the fatty acids that are β-oxidized, converted to β-HB, and contribute to ACoA. The intermediate metabolite ACoA is a product of pyruvate and OS fatty acids and is the entry point of the citric acid cycle, also known as the Krebs cycle or tricarboxylic acid (TCA) cycle. Citrate provides a self-regulating mechanism through its inhibition of PFK. If citrate builds up, it signals the cell that the citric acid cycle is backed up and does not need more intermediates to create ATP, slowing down glycolysis. This in turn reduces the production of pyruvate and lactate. The six key metabolic processes under consideration in this study are described by equations (1)–(6) and the 25 parameters, following key features of photoreceptor biochemistry [10, 29, 31]. As such, we define glycerol-3-phosphate as

Table 2 Parameter descriptions and units

Parameter	Description	Units
λ	Transport conversion factor	mM^{-1}
$V_{[G]}$	Maximum transport rate of glucose	$mM \cdot min^{-1}$
$K_{[G]}$	Substrate concentration that gives half the maximal rate of $V_{[G]}$	mM
n	Rate of passive glucose transport in the absence of RdCVF	$mM \cdot min^{-1}$
δ	Scaling factor for contribution of RdCVF by rods	no units
q	Fraction of G converted into G3P	no units
$V_{[G3P]}$	Maximum production rate of G3P	$mM \cdot min^{-1}$
$K_{[G3P]}$	Substrate concentration that gives half the maximal rate of $V_{[G3P]}$	mM
$V_{[PYR]}$	Maximum production rate of PYR	$mM \cdot min^{-1}$
$K_{[PYR]}$	Substrate concentration that gives half the maximal rate of $V_{[PYR]}$	mM
β	Rate of CIT inhibition of G catabolism (multiplied by an appropriate conversion factor)	mM^{-1}
α	Rate of β-oxidation of ingested OS fatty acids (created from G3P) to generate the β-HB substrate for ACoA	min^{-1}
ρ	Fraction of PYR converted into ACoA	no units
$V_{[LACT]}$	Maximum production rate of LACT	$mM \cdot min^{-1}$
$K_{[LACT]}$	Substrate concentration that gives half the maximal rate of $V_{[LACT]}$	mM
$V_{[ACoA]}$	Maximum production rate of ACoA	$mM \cdot min^{-1}$
$K_{[ACoA]}$	Substrate concentration that gives half the maximal rate of $V_{[ACoA]}$	mM
ϕ	Rate of CIT converted to ATP	min^{-1}
$V_{[CIT]}$	Maximum production rate of CIT	$mM \cdot min^{-1}$
$K_{[CIT]}$	Substrate concentration that gives half the maximal rate of $V_{[CIT]}$	mM
ψ	Maximum velocity of lactate transport	min^{-1}
k	Measurement of binding affinity of lactate transporter	mM^{-1}
γ	Maximum velocity of lactate transport contributing to ACoA	min^{-1}
G_E	Concentration of glucose outside the cell	mM
L_E	Concentration of lactate outside the cell	mM

G3P, which should not be confused with glyceraldehyde-3-phosphate (abbreviated as GAP, G3P, and GA3P in some literature).

Equation (1) describes the rate of change with respect to time of the glucose concentration. It increases or decreases proportionally to bidirectional glucose transport and decreases by catalysis. The transport function of [G3P] [10]:

$$\lambda n + \lambda \frac{V_{[G]}(\delta[G3P])^2}{K_{[G]}^2 + (\delta[G3P])^2} \qquad (7)$$

accounts for the passive transport term (first term of Equation (7)) and the facilitated transport term (second term of Equation (7)). In passive transport, glucose crosses the membrane without activation and stimulation by the facilitated transporter GLUT1, while in facilitated transport, RdCVF stimulates the transport activity of

GLUT1 by triggering its tetramerization and accelerating the uptake of glucose [10]. The expression δ[G3P] accounts for the concentration of RdCVF synthesized by rod phothoreceptors since it is assumed that RdCVF concentration is in proportion to [G3P].[1]

We model q as the fraction of [G] that is converted into [G3P] and $1 - q$ as the remaining fraction of [G] that is converted into [PYR]. The metabolism of glucose into these two metabolites is inhibited by [CIT], where β is the rate of citrate inhibition of glucose catabolism.

Equation (2) describes the rate of change with respect to time of the G3P concentration. [G3P] increases with an influx of glucose, which is inhibited by citrate, and decreases by production of OS, which serves as a measurement of β-oxidation of ingested OS fatty acids that contribute to the production of ACoA. We are taking catabolism of α[G3P] as a proxy for OS fatty acids converted into ACoA.[2]

Equation (3) describes the rate of change with respect to time of the pyruvate concentration. [PYR] increases with an influx of glucose, which is inhibited by citrate, and decreases by its conversion into lactate and ACoA. The factor $(1 - \rho)$ accounts for the fraction of [PYR] converted into lactate while ρ accounts for the fraction of [PYR] converted into ACoA.

Equation (4) describes the rate of change with respect to time of the lactate concentration. [LACT] increases by conversion of pyruvate to lactate via aerobic glycolosis and increases or decreases by bidirectional lactate transport. The lactate transport rate is modeled with a logistic function as follows:

$$f([\text{LACT}]) = \frac{\psi}{1 + e^{-k([\text{LACT}]-L_E)}}, \tag{8}$$

where ψ is the maximum transport rate, L_E is the extracellular concentration of lactate, and k is the binding affinity of lactate transporters which corresponds to the steepness of the curve f.

Since L_E accounts for the lactate concentration outside of the cell, the gradient flux of lactate is from inside to outside of the cell when [LACT] $> L_E$, while the opposite gradient flow occurs when [LACT] $< L_E$. If external lactate is in abundance, then the transport rate out of the cell is very small, i.e.,

$$f([\text{LACT}]) \approx 0, \quad \text{when} \quad L_E \gg [\text{LACT}].$$

In other words, if lactate inside of the cell is scarce, relative to external lactate, then the transport of lactate out of the cell is a slow process. If the intracellular

---

[1]There are approximately 20 rods per each cone in the human retina (and 25 to one in mice retina). G3P leads to the production of lipids which result in new photoreceptor OS. Thus we take [G3P] as a proxy for rods with the appropriate scaling factor incorporated into δ, the scaling factor for contribution of RdCVF by rods.

[2]Since we are not considering the RPE, we will utilize α[G3P] as a proxy for the metabolite β-hydroxybutyrate produced by the PRE and utilized by the photoreceptor's TCA.

lactate concentration is much larger than the extracellular concentration, then lactate transport out of the cell is faster, i.e.,

$$f([LACT]) \approx \psi, \quad \text{when} \quad [LACT] \gg L_E.$$

Equation (5) describes the rate of change with respect to time of the ACoA concentration. [ACoA] increases by PYR leakage to the mitochondria and β-HB produced from OS fatty acids generated by G3P lipid synthesis. It also increases or decreases by bidirectional lactate transport and decreases by its conversion into citrate. The lactate transport rate is modeled with a logistic function as follows:

$$h([LACT]) = \frac{\gamma e^{-k([LACT]-L_E)}}{1 + e^{-k([LACT]-L_E)}}, \tag{9}$$

where γ is the maximum transport rate and k is the steepness of the curve $h([LACT])$.[3] The extracellular lactate that comes into the cell gets converted into pyruvate which is immediately shuttled to the mitochondria for OXPHOS, and there is no re-conversion of lactate. Mathematically, this means that we can directly model the gradient influx of external lactate into the mitochondria and the conversion of this lactate to ACoA with the transport rate $h([LACT])$. The conversion of L_E to ACoA is metabolically faster when external lactate is in abundance, i.e.,

$$h([LACT]) \approx \gamma, \quad \text{when} \quad L_E \gg [LACT].$$

However, when the L_E is scarce, its contribution to the production of ACoA is negligible, i.e.,

$$h([LACT]) \approx 0, \quad \text{when} \quad [LACT] \gg L_E.$$

Equation (6) describes the rate of change with respect to time of the citrate concentration. [CIT] increases by the conversion of ACoA into citrate and decreases by its conversion into other intermediate metabolites leading to the creation of ATP.

In our model every resulting product becomes the substrate in the next metabolic reaction, with the exception of citrate, the last metabolite in our sequence of metabolic reactions, and lactate, which is modulated by L_E. The metabolic conversion of the substrates [G], [G3P], [PYR], and [ACoA] into their respective products, given by the variables in equations (1)–(6), are modeled with Hill type functions:

$$\frac{V_m S^n}{K_d^n + S^n},$$

---

[3]By the inverse relation of the functions $f([LACT])$ and $h([LACT])$, the parameters k and L_E have the analogous meaning with respect to each function.

where V_m is the maximal velocity of the reaction, K_d is the dissociation constant (or equivalently concentration of the substrate at which the conversion rate achieves its half-maximum value) and $n = 1, 2$ is the Hill coefficient. This coefficient relates to the number of binding sites available in the enzyme. When there is cooperative binding, n is greater than one, illustrating higher binding affinity of the substrate to the enzyme [46]. We modeled allosteric regulation kinetics with $n = 2$, indicative of multiple binding sites and enzyme cooperation, which results in increased substrate conversion rates after the first binding event.

2.3 Parameter Values

All model parameters and their meaning are described in Table 2. We performed an extensive literature search to identify and justify parameter values and ranges used in the model; see Table 3. When human values were not available, we used animal values. Note that even through $V_{[G3P]}$ and $V_{[PYR]}$ have the same baseline values, their corresponding range values, used later for the sensitivity analysis, are different. When metabolic parameter values for retina cells were not available, we used values from brain, heart, liver, or muscle tissues. Cancer cells can also serve as a case study to investigate the predictive capabilities of our model, as they also exhibit the Warburg effect, converting glucose to lactate even in the presence of oxygen. Since both cancer and photoreceptor cells utilize aerobic glycolysis for metabolism and both are high energy demanding cells, we used cancer data to see how well our cone cell model extends to other aerobic glycolysis systems.

3 Numerical Results

3.1 Model Validation

With parameter values in empirical ranges, we first verified that the model predicts a temporal evolution comparable to that observed in data. To do this, we compared model simulations with results from an empirical study in cancer cells, which provided measurements of the intracellular concentrations of glucose, lactate, and pyruvate over a period of four hours [54]. Ying et al. [54] measured these concentrations at six time points (0, 0.5, 1, 2, 3, and 4 h). 4T1 (breast cancer line) cells were cultured in 10 mM of both glucose and lactate with a pH of 7.4. The concentrations of glucose, lactate, and pyruvate were measured using a spectrophotometer.[4] We averaged the experimental results and used the resulting

---

[4]The authors generously shared their data used to generate their Figure 1B for three cells for each experiment.

Table 3 Parameter values used in simulations for photoreceptor model and cancer cells

Parameter	Normal photoreceptor model		Cancer cells		Reference
	Range value	Baseline value	Range value	Baseline value	
λ	[0.062, 0.093]	0.0755	[0.054, 0.101]		[10]+ & Est
$V_{[G]}$	[0.1, 1.56]	1.2			[10, 54]
$K_{[G]}$	[5, 24.7]	19			[10, 54]+
n	[0.0007, 0.0013]	0.001			[10]+
δ	[45.5, 95]	65			[10]+
q	[0.04, 0.2]	0.18	[0.336, 0.624]	0.38	[39]+ & Est
$V_{[G3P]}$	[0.12, 0.18]	0.15			[10]
$K_{[G3P]}$	[0.02, 0.171]	0.143			[10]+
$V_{[PYR]}$	[0.0013, 0.3915]	0.15			[10, 54]
$K_{[PYR]}$	[0.05, 2.21]	1.7			[10, 54]+
β	[0.7, 1.3]	1	[0.56, 1.04]	0.8	[19]+ & Est
α	[0.002, 1]	0.2			[10]+
ρ	[0.04, 0.06]	0.05			[10, 16, 17, 21, 51]
$V_{[LACT]}$	[0.098, 0.33]	0.14			[10, 54]
$K_{[LACT]}$	[0.0875, 10]	0.125			[10, 54]+
$V_{[ACoA]}$	[0.105, 0.195]	0.15			[54]
$K_{[ACoA]}$	[0.005, 0.02]	0.02			[35]
ϕ	[0.7, 1.3]	1	[0.35, 0.65]	0.5	[4] & Est
$V_{[CIT]}$	[0.021, 0.039]	0.03	[0.07, 0.13]	0.1	[4, 35] & Est
$K_{[CIT]}$	[0.0046, 0.00702]	0.0054	[0.35, 0.65]	0.5	[35] & Est
ψ	[6.4, 9.6]	8			[10]
k	[7, 13]	10	[0.315, 0.585]	0.45	CS & Est
γ	[0.7, 1.3]	1	[0.0105, 0.0195]	0.015	[27] & Est
G_E	[5, 20]	11.5	[8.4, 15.6]	13	[10, 51, 54] & Est
L_E	[5, 22]	10	[15.4, 28.6]	22	[10, 54] & Est

CS refers to parameter values unique to this current study. Est refers to parameter values estimated for cancer baseline values from the cancer cell line data shown in Table 5. + refers to values estimated from non-retina organs or reactant concentrations. In the columns for cancer cells we only reported ranges and baseline values which are different than those for the normal photoreceptor model

Table 4 Time series averages of glucose, lactate, and pyruvate [54]

Time (hours)	Lactate (mM)	Glucose (mM)	Pyruvate (mM)
0	2.8	1.87	0.042
0.5	13.39	7.98	0.1014
1	21.78	10.49	0.1358
2	21.55	9.81	0.1360
3	21.38	9.62	0.1469
4	21.98	8.49	0.1362

data, given in Table 4, as a first step in validating our model for an aerobic glycolysis system.

To account for the distinct molecular dynamics and the increased proliferation rates specific to cancer cells, we considered slightly different values for ten parameters than those for a healthy cone cell. See Table 3 for the ten parameter values labeled as estimated. The different parameters showed that a cancer cell undergoes aerobic gycolysis in a more disorganized manner while the aerobic glycolysis process in a cone cell is more controlled. The different parameter values in cancer revealed less controlled lactate transport in and out of the cell, with significantly slower lactate transport contributing to ACoA, a faster pace of cell growth, a slightly higher glucose flux, lower ability to self-regulate glycolysis through citrate inhibition or less abundance of ATP, and less production of intermediate metabolites for ATP production by citrate. The differences in these mechanisms are defined by a much lower k value (0.45 versus 10) in the gating functions $h([LACT])$ and $f([LACT])$, which illustrates back flow and not a complete on-off gating mechanism of lactate exchange between the extracellular and intracellular space with a significantly smaller γ, velocity of lactate transport into the cell for ACoA production; a higher fraction q of glucose converted to G3P for lipid synthesis and cell growth, which confirms the rapid cancer cell division and growth; a slightly larger range of glucose transport conversion factor λ, indicating more glucose supply variability, including a higher demand for glucose; smaller citrate inhibition of glycolysis β, signifying less self-regulation or potentially less abundance of ATP; and smaller rate ϕ of converting citrate to ATP, illustrating a reduction in ATP. The lack of tight metabolic regulation in cancer was further shown by the two fold increase in external lactate L_E, and the faster metabolic reaction of [CIT], given by the value of $V_{[CIT]}$, and the larger $K_{[CIT]}$ substrate concentration that gives half the maximal rate of $V_{[CIT]}$.

The model simulations show a good fit with the data, with all three concentrations stabilizing to their steady states within a little over an hour, as shown in Fig. 3. Our model assumes a constant external glucose flow allowing for steady levels of [G] to be achieved while the experimental data comes from cultured cells leading to an eventual decay in [G]. Though there are many similarities in metabolism between cancer cells and photoreceptors in that both cell types exhibit the Warburg effect, retinal cell parameters differ. However, this qualitative match to data is a good

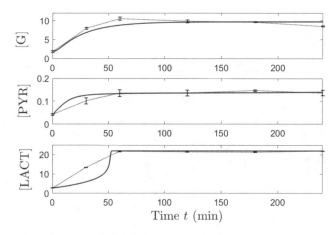

Fig. 3 Fitting model predictions with data from cancer cells. In each panel, the black curve shows the average empirical values, with error bars describing variability (standard deviation) over a population of three measured cells [54]. The blue curve represents our predicted solution. The parameters used are as follows: $\lambda = 0.0755$, $\rho = 0.05$, $\xi = 8$, $\delta = 65$, $\alpha = 0.2$, $\beta = 1$, $q = 0.38$, $n = 0.001$, $G_E = 13$, $k = 0.45$, $\gamma = 0.015$, $\phi = 0.5$, $L_E = 22$, $V_{[G]} = 1.2$, $K_{[G]} = 19$, $V_{[G3P]} = 0.15$, $K_{[G3P]} = 0.143$, $V_{[PYR]} = 0.15$, $K_{[PYR]} = 1.7$, $V_{[LACT]} = 0.14$, $K_{[LACT]} = 0.125$, $V_{[ACoA]} = 0.15$, $K_{[ACoA]}$ n 0.02, $V_{[CIT]} = 0.1$, $K_{[CIT]} = 0.5$. Initial conditions for the simulation were chosen to agree with the average empirical ones (in mM): [G] = 1.87; [G3P] = 0.12; [PYR] = 0.042; [LACT] = 2.8; [ACoA] = 0.03; [CIT] = 0.02

proof of concept for our model, which can now be tuned to parameters specific for photoreceptors.

3.2 Bifurcation Analysis and Bistability Ranges

As expected, the long-term dynamics of the system depend on its parameter values, and is altered by parameter perturbations. The model's sensitivity to changes and uncertainty in its parameters, which define various key mechanisms of the cone metobolic system, are further analyzed in Sect. 3.3. Here, we observe the effects of perturbing specific key parameters, and discuss the crucial consequences of the number and position of steady states (which correspond to specific physiological states and may distinguish between viability or failure of the system).

We first analyzed the changes in dynamics in response to variations in the external glucose concentration, G_E. Figure 4 shows the system's equilibria and their evolution and phase transitions as G_E is increased within the range of 0–13 mM. Each panel illustrates separately the projection of the same equilibrium curve along each of the variables in the system, representing key metabolite concentrations. The figure suggests that a reduced extracellular glucose supply below 2.6 mM (i.e., $G_E < 2.6$ mM) cannot successfully sustain the system and elevate internal glucose

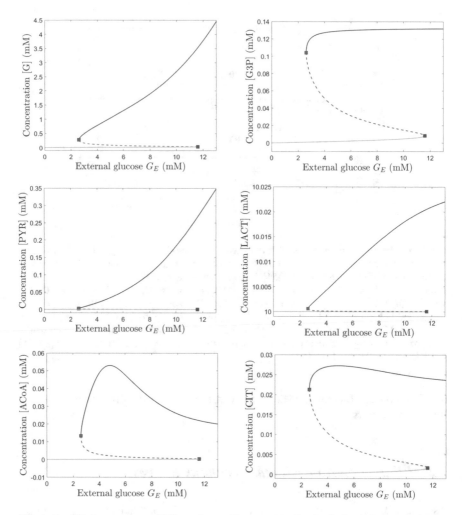

Fig. 4 Equilibrium curves and bifurcations with respect to G_E. As the level of external glucose is varied between G_E=0–13 mM, the equilibria of the system are plotted, each panel representing a different component of the same equilibrium curves. There are two locally stable equilibrium branches shown as green and blue solid curves, and a saddle equilibrium, shown as a dotted red curve. The bistability window onsets with a saddle node bifurcation at $G_E \sim 2.6$ mM (brown square marker), and closes with another saddle node bifurcation at $G_E \sim 11.6$ mM (purple square marker). The other system parameters were held fixed as: $\lambda = 0.0755$, $n = 0.001$, $\delta = 65$, $q = 0.18$, $\beta = 1$, α=0.2, $\rho = 0.05$, $\varphi = 1$, $\psi = 8$, $k = 10$, $\gamma = 1$, $L_E = 10$; $V_{[G]} = 1.2$, $K_{[G]} = 19$, $V_{[G3P]} = 0.15$, $K_{[G3P]} = 0.143$, $V_{[PYR]} = 0.15$, $K_{[PYR]} = 1.7$, $V_{[LACT]} = 0.14$, $K_{[LACT]} = 0.125$, $V_{[ACoA]} = 0.15$, $K_{[ACoA]} = 0.02$, $V_{[CIT]} = 0.03$, $K_{[CIT]} = 0.0054$

to a viable range. In this regime of $G_E < 2.6$ mM, the only attainable long-term physiological state, represented by the only stable equilibrium reachable from any initial conditions, is a "low functioning" stable state, shown as a green solid curve in the figure, and characterized by [LACT] ~10 mM with all other metabolite concentrations close to zero. This represents a pathological state of the cone.

At $G_E \sim 2.6$ mM, the system undergoes a saddle node bifurcation. If the external glucose level is raised past this phase transition value, the system enters a bistability regime, where a second, "viable" physiological steady state becomes available, with metabolite concentration levels in all components within a range for a healthy cone cell (illustrated in our panels as a blue solid branch of the equilibrium curve). Depending on the initial concentrations of the six metabolites in our model, the cone cell metabolism may converge to either the pathological or the healthy state. The [G], [G3P], [PYR], [LACT], [ACoA], and [CIT] levels change in response to G_E being further increased up to 13 mM. The internal glucose concentration [G] increases (up to ~4.5 mM), and so does the steady state level of [PYR], all the other components remain relatively unaltered, after a transient following up the birth of the second steady state. This shows the importance of external glucose and the components that alter it in driving the system via glucose and pyruvate metabolism.

The bistability window persists up to $G_E \sim 11.6$ mM, allowing different initial conditions to converge to one of two locally attracting equilibria (the green and the blue curves, separated by the unstable saddle shown as a red dotted curve). Convergence of different initial states in different attraction basins to either of the two stable steady states is further illustrated in Fig. 5. We show a [G]-[LACT] phase

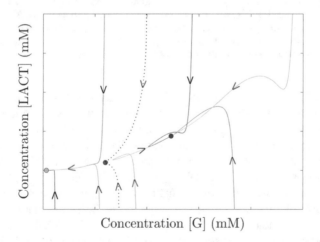

Fig. 5 Schematic representation of coexistence of equilibria within the bistability window, shown in a phase-space two-dimensional slice [G]-[LACT]. The two stable equilibria are shown as a green and a blue dot. A third, saddle equilibrium is shown as a red dot. A few simulated trajectories converging to the green equilibrium are shown as green curves, and simulated trajectories which converge to the blue equilibrium are shown as blue curves. The stable manifold of the saddle was symbolically drawn as a dotted black curve. The fixed parameters are the same as in Fig. 4. Figure 6 provides a more complete representation of all components for two representative solutions corresponding to two different initial conditions; one converging to the green dot, and one converging to the blue dot, for G_E=11.5 mM

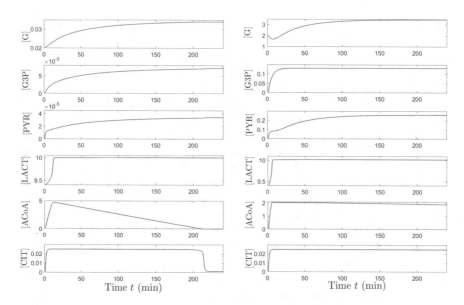

Fig. 6 Simulation of two solutions converging to the two different locally attracting equilibria, for our system in the bistability regime (with external glucose concentration G_E=11.5 mM and the rest of the parameters as in Figs. 4 and 5). The left versus the right panels represent two trajectories which differ only in their initial glucose concentration level. The other initial concentration are [G3P]=[PYR]=[ACoA]=[CIT]=0, and [LACT] = 9.4 mM. **Left.** Initial glucose level [G] = 0.02 mM. The system converges to a non-viable steady state in which all concentrations are close to zero, except for lactate. **Right.** Initial glucose level [G] = 2 mM. The system converges to a biologically viable/ healthy steady state as observed in empirical studies

space slice, for a value of G_E within the bistability range. For a more complete illustration, Fig. 6 shows two potential evolutions of the system in the bistability regime (for $G_E = 10$ mM). The left panel illustrates all components of the solution for a set of initial conditions in the basin of attraction of the green ("low functioning" or unhealthy) stable state, and the right panel for the blue ("high functioning" or healthy) steady state. The bistability window ends at $G_E \sim 11.6$ mM, and henceforth the healthy equilibrium remains the only attainable state in the long run. The basin of attraction provides a range for the initial concentration levels of our six metabolites that will drive the system to either the pathological or healthy state depending on the parameter values. Investigating how varying the parameters leads to one of these two states provides potential mechanisms that can be altered as potential therapies for improving cone vitality and sight.

Tracking the behavior of the system in response to varying the transport conversion factor λ, or the rate of passive glucose transport n, leads to very similar bifurcation diagrams, bistability windows, and variable ranges. Thus, they are not further illustrated here. Instead, we focus on α, the rate of β-oxidation of ingested OS fatty acids created from G3P. While there is a bistability regime that lives

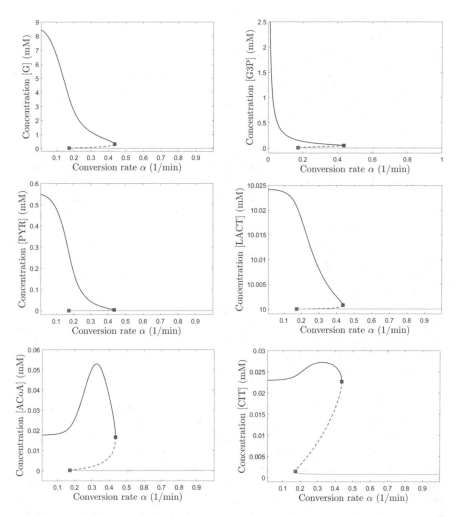

Fig. 7 Equilibrium curves and bifurcations with respect to α. As the rate α is varied between 0 and 1 min^{-1}, the equilibria of the system are plotted, each panel representing a different component of the same equilibrium curves. There are two locally stable equilibrium branches shown as green and blue solid curves, and a saddle equilibrium, shown as a dotted red curve. The bistability window onsets with a saddle node bifurcation at $\alpha \sim 0.17$ min^{-1} (brown square marker), and closes with another saddle node bifurcation at $\alpha \sim 0.43$ min^{-1} (purple square marker). G_E was fixed to 10 mM. The other system parameters were held fixed as in Fig. 4

between two saddle node points, the evolution of the system when varying α through these phase transitions is qualitatively different. We illustrate this behavior in Fig. 7.

When the rate of β-oxidation of fatty acids (α) exceeds the bifurcation value 0.43 min^{-1}, the system exhibits a unique locally stable equilibrium (solid green curve). This is a low functioning/unhealthy equilibrium, in the sense that all system components stabilize close to zero, except for [LACT] which stabilizes close to

10 mM, the external lactate value. Our analyses reveal that prolonged high rates of β-oxidation (beyond $0.43\,\text{min}^{-1}$) which exist under extreme glucose starvation and scarce key metabolites will result in the pathological unhealthy state without any alternative for reprogramming the cone to a healthy state by altering certain processes or mechanisms. Further, our findings show that the gating mechanisms of lactate transport in the cone cell is a tightly controlled mechanism and thus always stabilizes close to the external lactate concentration.

Varying α above the bistability regime does not have significant impact on the long term outcome. However, when α is reduced past the saddle node bifurcation (purple marker), the system suddenly enters its bistability regime and gains access to a second, high functioning/healthy steady state (blue solid curve). Rates of β-oxidation higher than $0.17\,\text{min}^{-1}$ and lower than $0.43\,\text{min}^{-1}$ provide the possibility of reprogramming the cone to a healthy state by altering certain processes and mechanisms. When α is decreased past the lower saddle node bifurcation at $\sim\!0.17\,\text{min}^{-1}$, the green curve disappears in the collision with the unstable equilibrium, and the high functioning steady state becomes the only stable long term outcome. This result confirms that low rates of β-oxidation are aligned with a robust healthy metabolic state for the cone that can not be perturbed.

Since the blue curve represents the healthy viable outcome, and in fact the only stable outcome for small enough values of α, it is useful to track its progression in response to perturbations of the parameter. As α is progressively lowered, there is first an increase in all steady state components of the system. After an initial upward and then downward transient, the [ACoA] and [CIT] concentrations will consistently settle to the same relatively low states ([ACoA] $\sim\!0.02\,\text{mM}$ and [CIT] $\sim\!0.025\,\text{mM}$) as α approaches zero. The other steady state components will continue to increase as α approaches zero. While [G], [PYR], and [LACT] still settle to values in the biological range, [G3P] exhibits a blowup as α approaches zero. This is not at all surprising, since the [G3P] concentration is the compartment affected most directly by the shutdown of pathway **k** (i.e., by reducing to zero the β-oxidation of ingested OS fatty acids created from G3P). Under starvation or additional need of energy, β-oxidation of fatty acids becomes a key substrate to fuel ATP production in the TCA cycle. The bifurcation analysis for α shows that when initial [G], [PYR], and [G3P] levels are relatively low, α has the ability to change the fate of the cone cell and its metabolism. But α can only do this within a small range of values. This shows that this process of creating energy via intermediate substrates created from β-oxidation of fatty acids is mainly an auxiliary process and the main process by which the cell relies on intermediate metabolites and substrates.

3.3 Sensitivity Analysis

We use sensitivity analysis to determine which processes have the greatest impact on the intracellular concentrations tracked by the model. Sensitivity analysis includes the following general steps: (i) vary the model parameters, (ii) perform model simulations, (iii) collect information on an output of interest (this can be the model output or another outcome), and (iv) calculate sensitivity measures. There are local and global sensitivity analysis methods. In local methods, parameters are varied one at a time, and in global methods, all parameters are varied at the same time. Examples of global sensitivity analysis methods include Latin Hypercube Sampling/Partial Rank Correlation Coefficient (LHS/PRCC), the Sobol method, and Extended Fourier Amplitude Sensitivity Test (eFAST). Depending on the technique used, sensitivity measures are called coefficients or indices, and they indicate the impact of parameter changes on the output of interest [34]. In LHS/PRCC, the sensitivity measure is named the partial rank correlation coefficient (PRCC). This method can only be applied when parameter variations result in monotonic changes in the output [7, 34]. However, the advantages of LHS/PRCC compared to other global methods are simplicity and much lower computational demand. The magnitude of the PRCC values provides information about parameter influence on the outcome of interest.

If the PRCC magnitude is greater than 0.4, the outcome of interest is considered sensitive to changes in the corresponding parameter [34]. The sign of a PRCC value shows if the corresponding parameter and the output are directly or inversely related. A positive coefficient indicates that the parameter and the output move in the same direction. A negative coefficient means they move in opposite directions, so as a parameter increases, the output decreases, and vice versa [7, 34]. In LHS/PRCC, parameters are varied simultaneously using Latin hypercube sampling (LHS). This involves assigning a probability distribution to each parameter, dividing the distribution into areas of equal probability and drawing at random and without replacement a value from every area [7, 34]. With LHS/PRCC, we can examine how a specific output is affected by an increase or decrease in a specific parameter, which can be useful for identifying the best parameters to target for control. Additionally, with LHS/PRCC we can explore how changes in initial conditions influence an outcome of interest [34].

The eFAST method can be conducted in the case when there are non-monotonic relationships between parameters (i.e., inputs) and a specific output of interest, but this approach is more computationally expensive than LHS/PRCC. In eFAST, the sensitivity measures are called sensitivity indices and they quantify the portion of variance in the outcome due to uncertainties in the parameters. There is a first order sensitivity index and total order sensitivity index. The first order index is a measure of how a parameter contributes to the output variance individually. The total order index shows the contribution a parameter makes to the output variance individually and in interaction with other parameters.

The magnitude of sensitivity indices determines the importance of parameters [34, 44, 45]. In eFAST, parameters are varied at the same time using a sinusoidal search curve, where angular frequency is specified for each parameter. To compute the sensitivity indices for a given parameter, a high frequency is assigned to that parameter, while all other parameters are assigned a low frequency [34, 44, 45]. With eFAST, we can examine which parameter uncertainties have the largest impact on output variability [34]. Due to the intricacies and complexity of eFAST, initial

conditions are rarely used as input factors. In the next section, we present the results
of our sensitivity analysis using both the LHS/PRCC and eFAST methods.

3.4 Sensitivity Results

The results of our sensitivity analysis are summarized in Table 5, which correspond
to the detailed results shown in Figs. 8, 9, and 10. For the normal photoreceptor
model, parameters are varied over their corresponding ranges given in Table 3. For
the case of cancer conditions, due to insufficient information regarding parameter
ranges, we allowed for 30% variation around nominal values.

3.4.1 Cancer Conditions

Both the LHS/PRCC and the eFAST sensitivity analysis results for cancer conditions show that the glucose level inside the cell, [G], is most sensitive to changes in
the parameter G_E, the concentration of glucose outside the cell. The two methods
also classify as important q (the fraction of glucose converted into G3P), δ (the
increased uptake of glucose facilitated by hypoxia inducible factor 1 signaling,
which up-regulates the expression of the glucose transporter GLUT1, for cancer
cells [24]), $V_{[G]}$ (the maximum transport rate of glucose), λ (the transport conversion
factor), α (the rate of β-oxidation of fatty acids [created from G3P]), and $K_{[G]}$ (the
substrate concentration giving half the maximal rate of $V_{[G]}$). These parameters are
involved in three key processes responsible for cell energy and growth; total glucose
uptake (a catalyst in both), the utilization of G3P in β-oxidation (that results in β-HB
which can be used as an oxidative substrate in the TCA cycle), and the production
of G3P for lipid synthesis (which is essential for growth). The PRCC results also
reveal that changes in the glucose concentration are inversely related with changes
in the parameters α and $K_{[G]}$; see Fig. 8.

The PRCC and eFAST analyses for cancer conditions both reveal that the
pyruvate concentration, [PYR], is sensitive to variation in the parameters which
capture the maximum production rate of lactate and pyruvate ($V_{[LACT]}$ and $V_{[PYR]}$,
respectively) and the fraction of glucose converted into G3P (q); see Table 5. The
negative PRCC values corresponding to the sensitivity of [PYR] to changes in q and
$V_{[LACT]}$ indicate that as the fraction of glucose diverted into G3P and the maximum
production rate of lactate decrease, the pyruvate concentration, [PYR], increases.
The PRCC approach also highlights how [PYR] is affected by $K_{[LACT]}$, which
measures the pyruvate concentration that gives half the maximal rate of $V_{[LACT]}$.
Changes in the chemical reaction of [LACT], in particular change in the mechanisms
within as defined by parameters $V_{[LACT]}$ and $K_{[LACT]}$, affect the resulting [PYR] levels.

According to both sensitivity analysis methods, the lactate concentration level
inside a cancer cell, [LACT], is significantly influenced by the concentration of
lactate outside the cell, L_E, both in its overall levels and its variability. The eFAST

Table 5 PRCC and eFAST results on the model applied to a cancer cell and a normal cone cell for $t_f = 240$ (minutes) with initial conditions [LACT] = 9.4, [G3P]=[PYR]=[ACoA]=[CIT]=0 for cone cell and [G] = 1.87, [G3P] = 0.12, [PYR] = 0.042, [LACT] = 2.8, [ACoA] = 0.03, and [CIT] = 0.02 for cancer conditions. The sign of the PRCC is shown in parentheses with a threshold for significance being 0.4 and $p < 0.01$, and both first order and total order sensitivity indices considered in determining significance for eFAST. Both are presented in decreasing (magnitude) order of significance with bold font representing |PRCC| ≥ 0.7 and comparably large magnitude for the parameters using eFAST

Sensitivity index	Parameters with significant sensitivity index for stated conditions	
(A) Uncertainty and sensitivity (US) analysis with [G] as outcome of interest over the different initial conditions		
	Normal with [G](0) = 0.02	Cancer conditions
eFAST	$\alpha, G_E, K_{[G]}, V_{[G]}, V_{[PYR]}$	$G_E, q, \alpha, \delta, \lambda, K_{[G]}, V_{[G]}$
PRCC	$K_{[G3P]}, G_E$	$\boldsymbol{G_E, \delta, \lambda, V_{[G]}, q, \alpha(-), K_{[G]}(-), V_{[G3P]}}$
(B) Uncertainty and sensitivity (US) analysis with [G3P] as outcome of interest over the different initial conditions		
	Normal with [G](0) = 0.02	Normal with [G](0) = 2	Cancer conditions
eFAST	α	$\alpha, G_E, K_{[G]}, V_{[G]}, V_{[PYR]}$	$\alpha, V_{[G3P]}, q$
PRCC	$\alpha(-), G_E, K_{[PYR]}$	$\alpha(-), q, V_{[G3P]}$	$\alpha(-), V_{[G3P]}, q$
(C) Uncertainty and sensitivity (US) analysis with [PYR] as outcome of interest over the different initial conditions		
	Normal with [G](0) = 0.02	Normal with [G](0) = 2	Cancer conditions
eFAST	$\alpha, V_{[PYR]}, K_{[G]}, G_E, V_{[G]}$	$\alpha, V_{[PYR]}$	$V_{[LACT]}, V_{[PYR]}, q$
PRCC	$G_E, K_{[PYR]}(-), K_{[G3P]}$	$\alpha(-), V_{[PYR]}, K_{[G]}(-), q$	$\boldsymbol{q(-), V_{[LACT]}(-), V_{[PYR]}, K_{[LACT]}}$
(D) Uncertainty and sensitivity (US) analysis with [LACT] as outcome of interest over the different initial conditions		
	Normal with [G](0) = 0.02	Normal with [G](0) = 2	Cancer conditions
eFAST	$L_E, \alpha, K_{[G]}$	$L_E, \alpha, K_{[G]}$	$L_E, k, V_{[PYR]}$
PRCC	$[LACT]_0$	L_E	L_E
(E) Uncertainty and sensitivity (US) analysis with [ACoA] as outcome of interest over the different initial conditions		
	Normal with [G](0) = 0.02	Normal with [G](0) = 2	Cancer conditions
eFAST	L_E, γ, α	L_E, α, γ	$L_E, k, q, \gamma, V_{[CIT]}$
PRCC	$L_E, [LACT]_0(-), \gamma$	$L_E, [LACT]_0(-)$	$\boldsymbol{L_E, k}, V_{[CIT]}(-), q, V_{[G3P]}$
(F) Uncertainty and sensitivity (US) analysis with [CIT] as outcome of interest over the different initial conditions		
	Normal with [G](0) = 0.02	Normal with [G](0) = 2	Cancer conditions
eFAST	$L_E, \phi, V_{[CIT]}$	$L_E, \phi, V_{[CIT]}$	$\phi, L_E, V_{[CIT]}, k, V_{[G3P]}, q$
PRCC	$\phi(-), V_{[CIT]}, L_E$	$\phi(-), V_{[CIT]}, L_E$	$\phi(-), L_E, V_{[CIT]}, q$

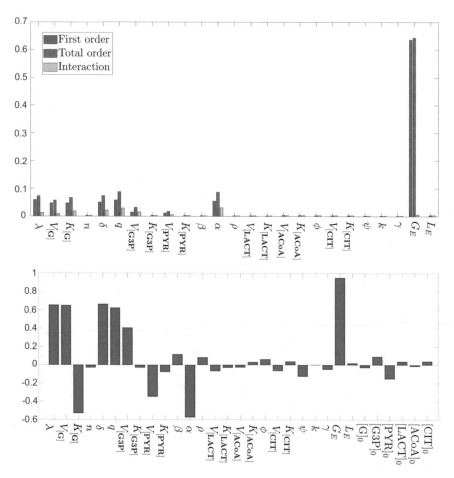

Fig. 8 Sensitivity results for [G] using eFAST (top) and LHS/PRCC (bottom) for cancer conditions. These are the graphical results that are entered in the Cancer conditions column of Table 5 for [G]. Both methods show that [G], the glucose level inside the cell, is most sensitive to changes in the parameter G_E. Inspection of the top and bottom graphs shows comparable relative impact on [G] for the remaining parameters, with the exception of $V_{[G3P]}$ which stands out as influential using PRCC but not eFAST

approach highlights two additional parameters that affect the variability of [LACT]. Uncertainty in the maximum production rate of pyruvate, $V_{[PYR]}$, and the binding affinity level of the lactate transporter, k, will result in the variability of [LACT]. The sensitivity results for [G], [PYR], and [LACT] indicate that the initial biochemical reactions in the gycolysis pathway are sensitive to more mechanisms, as illustrated by the number of parameters in the corresponding cases in the Cancer conditions column of Table 5, than the reactions further downstream not including the reactions in the TCA and the Kennedy pathways.

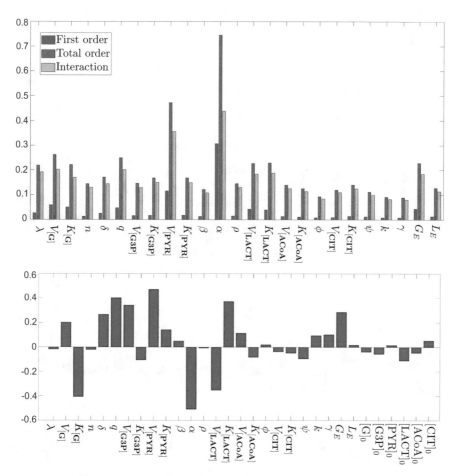

Fig. 9 **Sensitivity results for [PYR] using eFAST (top) and LHS/PRCC (bottom) using normal photoreceptor conditions with [G](0) = 2.** These are the graphical results that are entered in the relevant column of Table 5 for [PYR]. The graphs illustrate agreement in the importance of α and $V_{[\text{PYR}]}$

3.4.2 Normal Cone Cell

Both the LHS/PRCC and eFAST methods show that for a cone cell, where model simulations are performed with initial condition for glucose of [G](0) = 0.02, the glucose concentration is sensitive to changes in the level of external glucose (G_E). The results from the eFAST approach indicate that uncertainty in the rate of β-oxidation of ingested OS fatty acids (created from G3P) (α) has the more significant impact in the variability of [G]. For low initial concentration levels of glucose there are more mechanisms affecting [G] variability as indicated by the eFAST results. The LHS/PRCC and eFAST sensitivity results also have other differences. While

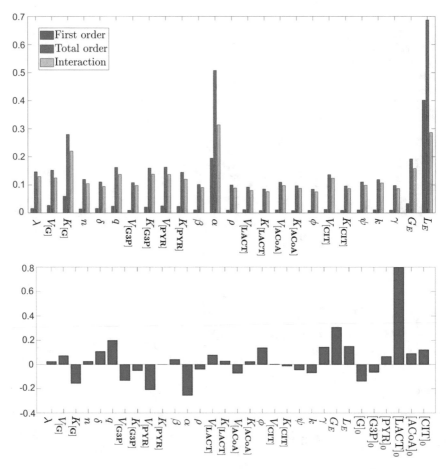

Fig. 10 **Sensitivity results for [LACT] using eFAST (top) and LHS/PRCC (bottom) using normal photoreceptor conditions with [G](0) = 0.02.** These are the graphical results that are entered in the relevant column of Table 5 for [LACT]. While eFAST classified L_E and α as most influential, PRCC determined that the initial lactate concentration is most important; see the text for further discussion of these results

PRCC highlights the substrate concentration that gives the half-maximum rate of $V_{[G3P]}$ as having an impact on the intracellular glucose concentration, this parameter is not classified as important by eFAST. On the other hand, eFAST indicates that [G] is sensitive to variation in the parameters $K_{[G]}$, $V_{[G]}$, and $V_{[PYR]}$.

The eFAST results using a higher initial condition for glucose of [G](0) = 2 show that the processes important for the intracellular glucose level are β-oxidation of OS fatty acids, external glucose, glucose uptake, and pyruvate production. In addition to indicating the impact of these factors, the PRCC method highlights the influence of converting glucose to G3P. For [G](0) = 2, the same parameters impact [G] in the eFAST results. However, for PRCC the number of mechanisms affecting [G] increases, so now changes in seven parameters (as opposed to two) affect [G]. These parameters are involved in glucose uptake, [PYR] biochemical reaction, glucose diversion to G3P, and β-oxidation. The negative sign of α in the PRCC analysis indicates that the glucose level in a cone cell decreases as the cell breaks down OS fatty acids at a higher rate to synthesize β-HB to be utilized as a substrate in the production of ACoA.

According to the eFAST results with [G](0) = 0.02, the pyruvate concentration, [PYR], in a cone cell is influenced by the oxidation of OS fatty acids (α), the amount of glucose outside the cell (G_E), the half limiting value of the glucose transport rate ($K_{[G]}$), the maximum rate of glucose transport ($V_{[G]}$), and the maximum production rate of pyruvate ($V_{[PYR]}$). In PRCC only three parameters affect [PYR]; G_E, $K_{[PYR]}$, and $K_{[G3P]}$. The substrate that gives the half-maximal rate of $V_{[PYR]}$, defined by $K_{[PYR]}$, inversely affects [PYR]. An increase in $K_{[PYR]}$ will increase the amount of substrate required for [PYR] to reach its saturation level.

The eFAST and PRCC results with [G](0) = 2 differ from those with lower initial condition for glucose: eFAST no longer classifies external glucose and glucose uptake as influential, and PRCC shows a whole new set of parameters as being important. In addition, there are less mechanisms (defined by the model's parameters) affecting [PYR] in the eFAST results as compared with the LHS/PRCC results for [G](0) = 2. Both sensitivity methods highlight [PYR] as being affected by changes and uncertainties in the parameters that describe β-oxidation and maximum production rate of [PYR], α and $V_{[PYR]}$, respectively. The PRCC results indicate an inverse relationship between variation in α and $K_{[G]}$ and changes in [PYR]. PRCC also identifies the process of diverting glucose to G3P for production of OS, which are rich in fatty acids, as having a strong effect. See Fig. 9 for the case with initial condition [G](0) = 2.

When the intracellular lactate level, [LACT], is the outcome of interest, both sensitivity analysis methods show as important the extracellular lactate level (L_E) when [G](0) = 2. The eFAST results also classify the processes of β-oxidation of OS fatty acids from lipids produced by G3P (α) and the half-limiting value of glucose transport ($K_{[G]}$) as having an impact on [LACT]. The eFAST results for [LACT] were the same for [G](0) = 0.02 and [G](0) = 2. The PRCC results show that only the lactate initial condition has an impact on the concentration of lactate inside the cell when the initial internal glucose concentration is low; see Fig. 10 for the case of with initial condition [G](0) = 0.02. The relatively small number of parameters that affect [LACT] levels and variability indicates the strong pull of these mechanisms (or parameters) to try to bring the external and internal lactate levels to a balance.

For both [G](0) = 0.02 and [G](0) = 2, PRCC and eFAST indicate that β-oxidation of OS fatty acids from lipids produced by G3P (α) is the most important process for the level and variability of [G3P]. In addition, PRCC shows that external glucose (G_E) and the half-limiting value of pyruvate production ($K_{[PYR]}$) have an impact on the [G3P] level for [G](0) = 0.02, while at the higher initial condition for glucose, important factors are the conversion of glucose to G3P (q) and the maximum production rate of G3P ($V_{[G3P]}$).

The eFAST results show that uncertainty in the same parameters influences the variability of [AcoA] at both $[G](0) = 0.02$ and $[G](0) = 2$. These parameters are L_E (external lactate level), α (β-oxidation of OS fatty acids), and γ (maximum velocity of lactate transport contributing to AcoA). PRCC also highlights γ as important but only for $[G](0) = 0.02$. With both initial conditions for glucose, the PRCC results indicate that the external lactate level and the initial internal lactate concentration have a strong impact on the level of ACoA.

The sensitivity results of [CIT] for $[G](0) = 0.02$ and $[G](0) = 2$ are the same for both methods. PRCC and eFAST indicate that [CIT] is impacted by external lactate (L_E), rate of CIT conversion to ATP (ϕ), and the maximum production rate of CIT ($V_{[CIT]}$). The relative impact of these mechanisms differs within each method. Uncertainties in external lactate affect the variability of [CIT] the most in eFAST, with the rate of CIT conversion to ATP having the second largest impact. PRCC reveals that ϕ affects [CIT] the most, with $V_{[CIT]}$ having the second largest impact. An increase in the rate of CIT conversion to ATP reduces the concentration of CIT, while an increase in the maximum production rate of CIT elevates [CIT].

Interestingly, across both the normal photoreceptor model and cancer conditions, changes in α affect [G] and [G3P] but only [PYR] in the cone cell. In an analogous manner, changes in external lactate, L_E, affect [LACT], [ACoA], and [CIT] across photoreceptor and cancer conditions (with one exception).

4 Discussion

4.1 Specific Comments on the Model

In this work, we developed and explored a mathematical model for the dynamics in the metabolic pathways of a healthy photoreceptor cell. We validated our model structure by comparing its predictions for concentrations of glucose, lactate, and pyruvate to data collected in cancer cells [54], which are metabolically similar to photoreceptors. In addition to developing the model structure, we also identified parameter values and ranges through a comprehensive literature search. When possible, we used values specific to photoreceptor cells and, if no measurements existed, selected another cell type as a proxy.

We applied two different global sensitivity analysis methods (LHS/PRCC and eFAST) and found the sensitive parameters resulting from each. PRCC reveals how the output of a model is affected if a parameter is changed, whereas variance-based methods such as eFAST measure which parameter uncertainty has the largest impact on output variability [34]. Using these two sensitivity analysis approaches in unison, we obtained a comprehensive view for which processes reflected in the equations (via the parameters) have the greatest impact on the metabolic system.

This sensitivity analysis, for the case of a photoreceptor cell, revealed that external glucose (reflected by G_E) and β-oxidation of fatty acids from OS (generated by G3P lipid synthesis) for ACoA production (defined by α) significantly affect the concentrations of glucose, G3P, and pyruvate at steady state (corresponding to time equal to 240 minutes). The importance of external glucose indicates that the effective metabolism of photoreceptors relies on sufficient availability of glucose, their primary fuel resource. The influence of β-oxidation of fatty acids, which links glycolysis occurring in the cytosol to oxidative phosphorylation in the mitochondria, suggests that photoreceptor metabolism is modulated by this feedback mechanism. The PRCC results also indicate that at a low initial level of intracellular glucose, the pyruvate concentration is most sensitive to changes in external glucose, while if greater amount of intracellular glucose is present initially, the pyruvate concentration is most sensitive to changes in the rate of β-oxidation.

Our sensitivity analyses reveal that ACoA, CIT, and intracellular lactate are impacted to the greatest extent by external lactate (L_E). Furthermore, they are the only metabolites sensitive to external lactate. This suggests that external lactate is an important mechanism affecting oxidative phosphorylation, while it does not seem to have a strong influence on the glycolysis pathway, where external glucose has a crucial role.

We used bifurcation techniques to study the dependence of the system's behavior on the parameters, in particular on G_E and α, identified as key parameters by the sensitivity analysis. We found that the system undergoes two saddle node bifurcations with respect to these parameters revealing bistability over a range of parameter values. This is heartening, as a properly designed analysis should reveal bifurcation parameters to be sensitive [34]. Bistability allowed us to investigate the mechanisms, defined by the parameters, that can be altered to bring a cone to healthy conditions from the pathological state. We were also able to determine key ranges for G_E and α as well as initial metabolite levels that will lead to one state versus another with the aid of bifurcation curves and basins of attraction.

Our analysis found that the system behaves monotonically (broadly speaking) as the external glucose concentration is increased. This is not surprising when considering the molecular coupling: as more glucose is made available to the cell, the internal glucose concentrations are expected to increase, driving in turn (via pathways **c** and **d**) higher concentrations of G3P and PYR, respectively, and further (via pathway **g**) a higher concentration of LACT. The level of [LACT] eventually increase above the external concentration L_E, preventing additional production of [ACoA], and subsequently of [CIT]. Our bifurcation analysis also reveals that a very low external source of glucose (less than 2.6 mM) cannot drive the cell to function in a healthy regime, since in the long term all metabolites will be depleted without an adequate source of glucose to maintain cone metabolism, except for the β-HB and external lactate used as substrates for ATP production.

Increasing G_E to an adequate level pushes the system into the bistability window, with two potential, and very different outcomes. This opens up the possibility for the cell to function in a healthy long-term regime (with concentrations which have been observed empirically within the healthy functional range for the eye). This alternative prognosis is available based on the cone cell's current state or ability to change the current metabolite levels in the cell. If all the molecular pools of the cell have become extremely low, the cell can no longer be rescued by increasing the extracellular glucose. Figure 4 illustrates how an external concentration $G_E =$ 11.5 mM or lower cannot resuscitate a cell with already depleted molecular pools [G3P]=[PYR]=[ACoA]=[CIT]=0, [LACT] close to the external concentration of 10 mM, and [G] = 0.02 mM. The same external concentration of $G_E = 11.5$ mM can bring a cell to a healthy regime provided the initial glucose is as high as [G] = 2 mM. Such a surge of glucose input may have, however, other physiological effects on the system, not captured by our model, and should not be necessarily viewed as a cure-all strategy.

Our sensitivity and bifurcation analyses support the expectation from the model diagram (Fig. 2) that the system's prognosis also depends quite crucially on the rate α of β-oxidation of ingested OS fatty acids. The metabolites directly affected by even small changes in α are [G], [G3P], [PYR], [LACT], and [ACoA], but these perturbations propagate via the tight coupling of the system, affecting the long-term concentrations of all its components.

Our bifurcation analysis in Sect. 3.2 shows the global effects of having an overly glucose-starved system corresponding to an extremely large rate of β-oxidation. Overall, too high of a β-oxidation rate leads to a complete system shutdown. An extremely low β-oxidation rate leads to a dangerously high accumulation of G3P in the cell, as the system under-utilizes lipids to be converted to OS (whose fatty acids will eventually be used to generate β-HB).

The bifurcation analysis for α between $\sim$0.17–0.43 mM, shows that the bistability regime, hence the optimal functioning of the system, is contingent on its initial state. If the current state of the cell is close to healthy, further tuning its β-oxidation rate (e.g., via medication or therapies) can optimize its function. However, if the cell's current state is very poor (e.g., based on a history of functioning under pathological parameter values), the behavior cannot be rescued even by a substantial adjustment in the β-oxidation rate, and the cell's function will remain poor. These effects support our sensitivity analysis, which showed all metabolites of the system (with the exception CIT) to be sensitive to changes or perturbation of α.

It has been established that the RPE serves as the principal pathway for the exchange of metabolites (in particular, glucose and lactate) between the choroidal blood supply and the retina [14]. The fact that external glucose (G_E) and the rate of β-oxidation of fatty acids (α) are highlighted as important by our sensitivity and bifurcation analyses, as well as external lactate (L_E) in the sensitivity analyses, points to the critical role the RPE plays in photoreceptor metabolism. As these processes link the metabolism of photoreceptors with the metabolism of the RPE, our findings indicate that the normal function of photoreceptors relies heavily on their interaction with the RPE. This aligns with the physiological understating that photoreceptors and the RPE have a reciprocal resource relation and operate as a functional unit: the RPE provides photoreceptors with a source of metabolism via glucose, and photoreceptors provide a source of metabolism for the RPE via lactate.

External lactate is key for maintaining a balanced reciprocal resource relation between the RPE and photoreceptors, on which cone nutrition and vitality depend. In addition to glucose supplied by the RPE, photoreceptors can also consume lactate, produced by other retinal cells for oxidative metabolism. For example, Müeller cells are known to secrete lactate which can be used as fuel in photoreceptors [49]. The high impact of external lactate seen in the sensitivity analysis also points to the importance of this mechanism for photoreceptor metabolism.

4.2 Limitations and Future Work

We have identified three limitations in our model. (1) Our model considers a single healthy photoreceptor cell, whereas in reality there are multiple photoreceptors of different types and other cell types such as RPE and Müeller cells, forming a "metabolic ecosystem." Future work will address this complexity. (2) Our model would be improved if time series data existed for concentration levels of the metabolites, represented by state variables, in the retina. We identified time series concentrations of glucose, lacate, and pyruvate for cancer cells, which are metabolically similar to photoreceptors, but we would expect parameters to differ. Time series data would also allow us to better estimate parameters to which our model outputs are sensitive. (3) Some parameter ranges were not available from the literature, so we were forced to use a different tissue type as a proxy. Because both LHS/PRCC and eFAST depend on starting with biologically relevant ranges for each parameter and sampling within that range, this is a concern. If photoreceptor-specific parameter ranges become available, this model could be updated and improved.

The RPE is a layer of cells which provides glucose to photoreceptors, and these cells are also metabolically active. External glucose is very important, and though it is a parameter in our single-cell model, in future work it will depend on the dynamics of other cell types. The RPE serves as the main pathway for the exchange of critical metabolites (specifically, glucose and lactate) between the choroidal blood supply and the retina [14]. Metabolites that can be used as substitutes for photoreceptor energy production, during glucose deprivation, are also mediated by the RPE. Müeller cells are known to secrete lactate which can be used as fuel in photoreceptors [49]. A future step in this work will be investigating the interaction of the RPE, Müeller cells, and photoreceptors along with the "metabolic ecosystem" they create.

5 Conclusions

We developed and analyzed a mathematical model for the dynamics in the metabolic pathways of a healthy photoreceptor cell. Using two different methods for sensitivity analysis, we identified the parameters and potential mechanisms that are driving system output levels and variability which are particularly relevant to photoreceptor health. The behavior of the model for different values of the highly sensitive parameters was explored, and we demonstrated that certain sets of parameters exhibit phase transitions and bistable behavior where healthy and pathological states both exist.

Our work confirms the necessity for the external glucose, β-oxidation, and external lactate concentrations, which are key feedback mechanisms connecting the RPE and photoreceptors, to sustain the cell. The role of β-oxidation of fatty acids which fuel oxidative phosphorylation under glucose- and lactate-depleted conditions, is validated. A low rate of β-oxidation corresponded with the healthy cone metabolite concentrations in our simulations and bifurcation analysis. Our results also show the modulating effect of the lactate differential (internal versus external) in bringing the system to steady state; the bigger the difference, the longer the system takes to achieve steady state. Additionally, our parameter estimation results demonstrate the importance of rerouting glucose and other intermediate metabolites to produce glycerol 3-phosphate (G3P), to increasing lipid synthesis (a precursor to fatty acid production) to support the cone cell high growth rate. A number of parameters are found to be significant; however, the rate of β-oxidation of ingested outer segments is shown to consistently play an important role in the concentration of glucose, G3P, and pyruvate, whereas the extracellular lactate level is shown to consistently play an important role in the concentration of lactate and acetyl coenzyme A.

These mechanisms can be posed to the biology community for future experiments or for potential therapeutic targets. The ability of these mechanisms to affect key metabolites' variability and levels (as revealed in our analyses) signifies the importance of inter-dependent and inter-connected feedback processes modulated by and affecting both the RPE's and cone's metabolism. The modeling and analysis in this work provide the foundation for a more biologically complex model that metabolically couples different cell types as found in the retina.

Acknowledgments This work was initiated during the Association for Women in Mathematics collaborative workshop Women Advancing Mathematical Biology hosted by the Institute of Pure and Applied Mathematics at University of California, Los Angeles in June 2019. Funding for the workshop was provided by IPAM, NSF ADVANCE "Career Advancement for Women Through Research-Focused Networks" (NSF-HRD 1500481). A. Dobreva also received support from National Science Foundation Grant RTG/DMS-1246991. A. Radulescu is also supported by a Simons Foundation Collaboration Grant for Mathematicians. We are grateful to Nancy Philp for introducing us to the importance of lactate consumption in photoreceptor metabolism and for fruitful discussions related to this work. We would like to thank the two anonymous reviewers for their thoughtful comments that made the manuscript clearer and stronger.

Appendix

A   Simple Model

In this section, System (1)–(6) is simplified to gain more insight into the cone glucose metabolic pathways. The model is simplified as below:

$$\frac{d[G]}{dt} = \lambda(G_E - [G])\left(n + V_{[G]}\delta[G3P]\right) - \left(qV_{[G3P]} + (1-q)V_{[PYR]}\right)[G]$$

$$(10)$$

$$\frac{d[G3P]}{dt} = qV_{[G3P]}[G] - \alpha[G3P] \tag{11}$$

$$\frac{d[PYR]}{dt} = (1-q)V_{[PYR]}[G] - \left((1-\rho)V_{[LACT]} + \rho V_{[ACoA]}\right)[PYR] \tag{12}$$

$$\frac{d[LACT]}{dt} = (1-\rho)V_{[LACT]}[PYR] - \psi([LACT] - L_E) \tag{13}$$

$$\frac{d[ACoA]}{dt} = \rho V_{[ACoA]}[PYR] + \gamma(L_E - [LACT]) - V_{[CIT]}[ACoA] + \alpha[G3P]$$

$$(14)$$

$$\frac{d[CIT]}{dt} = V_{[CIT]}[ACoA] - \phi[CIT]. \tag{15}$$

We were able to prove existence and uniqueness of the model solution in System (10)–(15). Furthermore, it was proved that the system evolves to a unique equilibrium point under healthy conditions. However, the simple model does not capture the complete qualitative behaviours of the full model as shown by the stability analysis of the simple model.

Note that System (10)–(15) is well-posed and that all solutions remain within the state space, $[G] \geq 0$, $[G3P] \geq 0$, $[PYR] \geq 0$, $[LACT] \geq 0$, $[ACoA] \geq 0$, $[CIT] \geq 0$, since the right-hand side functions of System (10)–(15) are continuously differentiable [48]. The analysis of Model (10)–(15) is done by finding the equilibria and their corresponding stability properties. Setting the right-hand sides of the equations (10)–(15) equal to zero yields the following biological meaningful equilibrium point denoted by $E([G^*], [G3P^*], [PYR^*], [LACT^*], [ACoA^*], [CIT^*])$, and defined as follows

$$[G^*] = G_E - \frac{1}{2}\left(G_E + \frac{a}{\lambda b} + \frac{n}{b} - \sqrt{\frac{n^2}{b^2} + (G_E - \frac{a}{\lambda b})^2 + \frac{2n}{b}(G_E + \frac{a}{\lambda b})}\right)$$

$$(16)$$

$$[G3P^*] = \frac{q V_{[G3P]}}{\alpha}[G^*] \tag{17}$$

$$[PRY^*] = \frac{(1-q)V_{[PYR]}}{(1-\rho)V_{[LACT]} + \rho V_{[ACoA]}}[G^*] \tag{18}$$

$$[LACT^*] = L_E + \frac{(1-q)(1-\rho)V_{[LACT]}V_{[PYR]}}{\psi((1-\rho)V_{[LACT]} + \rho V_{[ACoA]})}[G^*] \tag{19}$$

$$[ACoA^*] = \frac{\kappa}{V_{[CIT]}}[G^*] \tag{20}$$

$$[CIT^*] = \frac{\kappa}{\phi}[G^*], \tag{21}$$

where $a = q V_{[G3P]} + (1-q)V_{[PYR]}$, $b = q\delta V_{[G]}V_{[G3P]}/\alpha$ and

$$\kappa = q V_{[G3P]} + \left[\rho V_{[ACoA]} - \frac{\gamma(1-\rho)V_{[LACT]}}{\psi}\right]\frac{(1-q)V_{[PYR]}}{(1-\rho)V_{[LACT]} + \rho V_{[ACoA]}},$$

with $\kappa \geq 0$.

Theorem 1 *The equilibrium E exists and it is locally stable.*

Proof From the parameter modeling assumptions, it is easy to prove that $[G^*] > 0$. Therefore, Equations (17)–(19) are all positive and Equations (20)–(21) are non-negative if and only if $\kappa \geq 0$. Hence, E is a biologically feasible equilibrium, since all the elements of E are non-negative for all parameter values of the model. Next, the Jacobian matrix corresponding to E is given by the following lower triangular block matrix:

$$J(E) = \begin{pmatrix} J_1(E) & \mathbf{0} \\ J_1^* & J_2(E) \end{pmatrix},$$

where J_1^* is a non-zero matrix and

$$J_1(E) = \begin{pmatrix} -\lambda(n + \delta V_{[G]}[G3P^*]) - a & \delta\lambda V_{[G]}(G_E - [G^*]) \\ q V_{[G3P]} & -\alpha \end{pmatrix},$$

$$J_2(E) = \begin{pmatrix} -(1-\rho)V_{[LACT]} - \rho V_{[ACoA]} & 0 & 0 & 0 \\ (1-\rho)V_{[LACT]} & -\psi & 0 & 0 \\ \rho V_{[ACoA]} & -\gamma & -V_{[CIT]} & 0 \\ 0 & 0 & V_{[CIT]} & -\phi \end{pmatrix}.$$

Therefore, the eigenvalues of $J(E)$ are determined from the eigenvalues of $J_1(E)$ and $J_2(E)$. Since $J_2(E)$ is a lower triangular matrix, its eigenvalues are given by its diagonal elements, which are all negative by the parameter modeling assumptions. From Routh-Hurwitz criteria, $n = 2$, the eigenvalues of $J_1(E)$ are negative or have negative real part if and only if $det(J_1(E)) > 0$ and $tr(J_1(E)) < 0$. From the model assumptions follow that $tr(J_1(E)) = -\lambda(n + \delta V_{[G]}[G3P^*]) - a - \alpha < 0$ and

$$det(J_1(E)) = \alpha\left(\lambda(n + \delta V_{[G]}[G3P^*]) + a\right) + \delta\lambda q\, V_{[G]} V_{[G3P]}([G^*] - G_E) > 0$$

if an only if

$$[G^*] > \frac{1}{2}\left(G_E - \frac{a}{b\lambda} - \frac{n}{b}\right).$$

From Equation (16)

$$[G^*] = \frac{1}{2}\left(G_E - \frac{a}{b\lambda} - \frac{n}{b}\right) + \sqrt{\frac{n^2}{b^2} + (G_E - \frac{a}{\lambda b})^2 + \frac{2n}{b}(G_E + \frac{a}{\lambda b})}$$
$$> \frac{1}{2}\left(G_E - \frac{a}{b\lambda} - \frac{n}{b}\right).$$

Therefore, all the eigenvalues of $J(E)$ are negative and hence E is a locally stable node.   □

Therefore in a long term glucose metabolic dynamic behaviour within a single cone, all the substrate variables achieve steady-state values, which depend linearly on the steady-state glucose concentration value, $[G^*]$, Equation (17)–(21). Furthermore, the concentration of glucose inside of the cell is always less than the outside concentration while the lactate concentration inside is more than the outside concentration, i.e.,

$$0 < [G^*] < G_E \quad \text{and} \quad 0 < L_E < [LACT^*]. \tag{22}$$

Note that $[G^*] = 0$ if and only if $\alpha = 0$ or $n = 0$ and $0 < G_E < a/b\lambda$. Therefore α and n are important parameters for the survival of the cell. Another important parameter is κ, since when $\kappa = 0$ the $[ACoA^*] = 0$, and $[CIT^*] = 0$ which also leads to a pathological metabolic steady-state outcome.

Figure 11 shows the molecular evolution of System (10)–(15), where the variables evolve to their steady-state values in about 3.3 h (200 min).

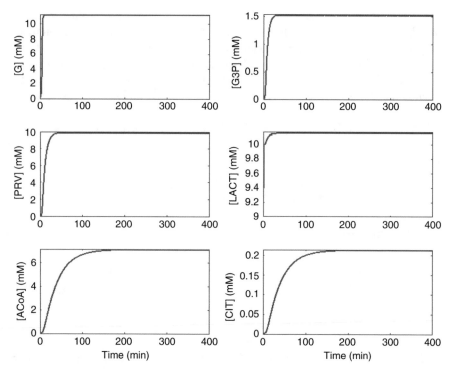

Fig. 11 Molecular evolution of energy demand in a single cone cell. The parameter values are set to their baseline value for normal photoreceptor model with initial conditions: [G] = 0.2, [LACT] = 9.4 and the rest of the other initial conditions are set to be equal to zero

B   Acronym Glossary

Acronym Definitions	
Acronym	Meaning
ACoA	acetyl coenzyme A
ATP	adenosine triphosphate
BSG-1	basigin-1
CIT	citrate
DHAP	dihydroxyacetone phosphate
eFAST	Extended Fourier Amplitude Sensitivity Test
G	glucose
GAP	glyceraldehyde-3-phosphate
G3P	glycerol-3-phosphate
G6P	glycerol-6-phosphate

(continued)

Acronym Definitions	
Acronym	Meaning
GLUT1	glucosetransporter 1
LACT	lactate
LDH	lactate dehydrogenase
LDHA	lactate dehydrogenase A
LDHB	lactate dehydrogenase B
LHS	Latin Hypercube Sampling
MCT	monocarboxylate transport proteins
NAD	nicotinamide adenine dinucleotide
OS	outer segment
OXPHOS	oxidative phosphorylation
PDC	pyruvate dehydrogenase complex
PFK	phosphofructokinase
PRCC	Partial Rank Correlation Coefficient
PYR	pyruvate
RdCVF	rod-derived cone viability factor
RdCVFL	rod-derived cone viability factor long form
RPE	retinal pigmented epithelium
TCA	tricarboxylic acid
VEGF	vascular endothelial growth factor

References

1. Jeffrey Adijanto, Jianhai Du, Cynthia Moffat, Erin L Seifert, James B Hurley, and Nancy J Philp. The retinal pigment epithelium utilizes fatty acids for ketogenesis implications for metabolic coupling with the outer retina. *Journal of Biological Chemistry*, 289(30):20570–20582, 2014.

2. Michalis Agathocleous, Nicola K Love, Owen Randlett, Julia J Harris, Jinyue Liu, Andrew J Murray, and William A Harris. Metabolic differentiation in the embryonic retina. *Nature cell biology*, 14(8):859, 2012.

3. Najate Aït-Ali, Ram Fridlich, Géraldine Millet-Puel, Emmanuelle Clérin, François Delalande, Céline Jaillard, Frédéric Blond, Ludivine Perrocheau, Sacha Reichman, Leah C Byrne, et al. Rod-derived cone viability factor promotes cone survival by stimulating aerobic glycolysis. *Cell*, 161(4):817–832, 2015.

4. Tiago C Alves, Rebecca L Pongratz, Xiaojian Zhao, Orlando Yarborough, Sam Sereda, Orian Shirihai, Gary W Cline, Graeme Mason, and Richard G Kibbey. Integrated, step-wise, mass-isotopomeric flux analysis of the tca cycle. *Cell metabolism*, 22(5):936–947, 2015.

5. Megha Barot, Mitan R Gokulgandhi, Vibhuti Agrahari, Dhananjay Pal, and Ashim K Mitra. Monocarboxylate transporter mediated uptake of moxifloxacin on human retinal pigmented epithelium cells. *Journal of Pharmacy and Pharmacology*, 66(4):574–583, 2014.

6. JM Berg. Tymoczko jl stryer l (2002). *Biochemistry. New York, NY: WH Freeman.*

7. S. M. Blower and H. Dowlatabadi. Sensitivity and uncertainty analysis of complex models of disease transmission: an HIV model, as an example. *International Statistical Review*, 62(2):229–243, 1994.

8. George A Brooks. The science and translation of lactate shuttle theory. *Cell metabolism*, 27(4):757–785, 2018.

9. Bang V Bui, Michael Kalloniatis, and Algis J Vingrys. Retinal function loss after monocarboxylate transport inhibition. *Investigative ophthalmology & visual science*, 45(2):584–593, 2004.

10. Erika T Camacho, Danielle Brager, Ghizlane Elachouri, Tatyana Korneyeva, Géraldine Millet-Puel, José-Alain Sahel, and Thierry Léveillard. A mathematical analysis of aerobic glycolysis triggered by glucose uptake in cones. *Scientific reports*, 9(1):4162, 2019.

11. Erika T Camacho, Claudio Punzo, and Stephen A Wirkus. Quantifying the metabolic contribution to photoreceptor death in retinitis pigmentosa via a mathematical model. *Journal of Theoretical Biology*, 408:75–87, 2016.

12. Erika T. Camacho and Stephen Wirkus. Tracing the progression of retinitis pigmentosa via photoreceptor interactions. *Journal of Theoretical Biology*, 317C:105–118, 2013.

13. Andrei O Chertov, Lars Holzhausen, Iok Teng Kuok, Drew Couron, Ed Parker, Jonathan D Linton, Martin Sadilek, Ian R Sweet, and James B Hurley. Roles of glucose in photoreceptor survival. *Journal of Biological Chemistry*, 286(40):34700–34711, 2011.

14. Víctor Coffe, Raymundo C Carbajal, and Rocío Salceda. Glucose metabolism in rat retinal pigment epithelium. *Neurochemical research*, 31(1):103–108, 2006.

15. Christine A Curcio, Kenneth R Sloan, Robert E Kalina, and Anita E Hendrickson. Human photoreceptor topography. *Journal of comparative neurology*, 292(4):497–523, 1990.

16. Jianhai Du, Whitney Cleghorn, Laura Contreras, Jonathan D Linton, Guy C-K Chan, Andrei O Chertov, Takeyori Saheki, Viren Govindaraju, Martin Sadilek, Jorgina Satrústegui, et al. Cytosolic reducing power preserves glutamate in retina. *Proceedings of the National Academy of Sciences*, 110(46):18501–18506, 2013.

17. Jianhai Du, Aya Yanagida, Kaitlen Knight, Abbi L Engel, Anh Huan Vo, Connor Jankowski, Martin Sadilek, Megan A Manson, Aravind Ramakrishnan, James B Hurley, et al. Reductive carboxylation is a major metabolic pathway in the retinal pigment epithelium. *Proceedings of the National Academy of Sciences*, 113(51):14710–14715, 2016.

18. Olivier Feron. Pyruvate into lactate and back: from the warburg effect to symbiotic energy fuel exchange in cancer cells. *Radiotherapy and oncology*, 92(3):329–333, 2009.

19. PB Garland, PJ Randle, and EA Newsholme. Citrate as an intermediary in the inhibition of phosphofructokinase in rat heart muscle by fatty acids, ketone bodies, pyruvate, diabetes and starvation. *Nature*, 200(4902):169–170, 1963.

20. Stephen S Goldman. Gluconeogenesis in the amphibian retina. lactate is preferred to glutamate as the gluconeogenic precursor. *Biochemical Journal*, 254(2):359–365, 1988.

21. Allison Grenell, Yekai Wang, Michelle Yam, Aditi Swarup, Tanya L Dilan, Allison Hauer, Jonathan D Linton, Nancy J Philp, Elizabeth Gregor, Siyan Zhu, et al. Loss of mpc1 reprograms retinal metabolism to impair visual function. *Proceedings of the National Academy of Sciences*, 116(9):3530–3535, 2019.

22. Andrew P Halestrap and Marieangela C Wilson. The monocarboxylate transporter family–role and regulation. *IUBMB life*, 64(2):109–119, 2012.

23. Douglas Hanahan and Robert A Weinberg. Hallmarks of cancer: the next generation. *cell*, 144(5):646–674, 2011.

24. N. Hay. Reprogramming glucose metabolism in cancer: can it be exploited for cancer therapy? *Nature reviews. Cancer.*, 16(10):635–649., 2016.

25. Ken-ichi Hosoya, Tetsu Kondo, Masatoshi Tomi, Hitomi Takanaga, Sumio Ohtsuki, and Tetsuya Terasaki. Mct1-mediated transport of l-lactic acid at the inner blood–retinal barrier: a possible route for delivery of monocarboxylic acid drugs to the retina. *Pharmaceutical research*, 18(12):1669–1676, 2001.

26. Mark A Kanow, Michelle M Giarmarco, Connor SR Jankowski, Kristine Tsantilas, Abbi L Engel, Jianhai Du, Jonathan D Linton, Christopher C Farnsworth, Stephanie R Sloat, Austin

Rountree, et al. Biochemical adaptations of the retina and retinal pigment epithelium support a metabolic ecosystem in the vertebrate eye. *Elife*, 6:e28899, 2017.

27. Emily Kenyon, Kefu Yu, Morten La Cour, and Sheldon S Miller. Lactate transport mechanisms at apical and basolateral membranes of bovine retinal pigment epithelium. *American Journal of Physiology-Cell Physiology*, 267(6):C1561–C1573, 1994.

28. Thierry Léveillard and Najate Aït-Ali. Cell signaling with extracellular thioredoxin and thioredoxin-like proteins: Insight into their mechanisms of action. *Oxidative medicine and cellular longevity*, 2017, 2017.

29. Thierry Léveillard, Nancy J Philp, and Florian Sennlaub. Is retinal metabolic dysfunction at the center of the pathogenesis of age-related macular degeneration? *International Journal of Molecular Sciences*, 20(3):762, 2019.

30. Thierry Léveillard and José-Alain Sahel. Rod-derived cone viability factor for treating blinding diseases: from clinic to redox signaling. *Science Translational Medicine*, 2(26):26ps16–26ps16, 2010.

31. Thierry Léveillard and José-Alain Sahel. Metabolic and redox signaling in the retina. *Cellular and Molecular Life Sciences*, 74(20):3649–3665, 2017.

32. Maria V Liberti and Jason W Locasale. The warburg effect: how does it benefit cancer cells? *Trends in biochemical sciences*, 41(3):211–218, 2016.

33. Gregory J Mantych, Gregory S Hageman, and Sherin U Devaskar. Characterization of glucose transporter isoforms in the adult and developing human eye. *Endocrinology*, 133(2):600–607, 1993.

34. S. Marino, I. B. Hogue, C. J. Ray, and D. E. Kirschner. A methodology for performing global uncertainty and sensitivity analysis in systems biology. *Journal of Theoretical Biology*, 254(1):178–196, 2008.

35. Yoichi Matsuoka and Paul A Srere. Kinetic studies of citrate synthase from rat kidney and rat brain. *Journal of Biological Chemistry*, 248(23):8022–8030, 1973.

36. Kyle S McCommis and Brian N Finck. Mitochondrial pyruvate transport: a historical perspective and future research directions. *Biochemical journal*, 466(3):443–454, 2015.

37. Daniel S Narayan, Glyn Chidlow, John PM Wood, and Robert J Casson. Glucose metabolism in mammalian photoreceptor inner and outer segments. *Clinical & experimental ophthalmology*, 45(7):730–741, 2017.

38. Haruhisa Okawa, Alapakkam P Sampath, Simon B Laughlin, and Gordon L Fain. Atp consumption by mammalian rod photoreceptors in darkness and in light. *Current Biology*, 18(24):1917–1921, 2008.

39. Mohammad Shamsul Ola and Kathryn F LaNoue. Molecular basis for increased lactate formation in the müller glial cells of retina. *Brain research bulletin*, 144:158–163, 2019.

40. GJ Pascuzzo, JE Johnson, and EL Pautler. Glucose transport in isolated mammalian pigment epithelium. *Experimental eye research*, 30(1):53–58, 1980.

41. Carol Lynn Poitry-Yamate, Serge Poitry, and Marcos Tsacopoulos. Lactate released by muller glial cells is metabolized by photoreceptors from mammalian retina. *Journal of Neuroscience*, 15(7):5179–5191, 1995.

42. Leslie A Real. The kinetics of functional response. *The American Naturalist*, 111(978):289–300, 1977.

43. Jane B Reece, Lisa A Urry, Michael Lee Cain, Steven Alexander Wasserman, Peter V Minorsky, Robert B Jackson, et al. *Campbell biology*. Number s 1309. Pearson Boston, 2014.

44. A. Saltelli, S. Tarantola, and F. Campolongo. Sensitivity analysis as an ingredient of modeling. *Statistical Science*, 15(4):377–395, 2000.

45. A .Saltelli, S. Tarantola, and K. P.-S. Chan. A quantitative model-independent method for global sensitivity analysis of model output. *Technometrics*, 41(1):39–56, 1999.

46. Herbert M Sauro. *Enzyme kinetics for systems biology*. Future Skill Software, 2011.

47. Lei Shi and Benjamin P Tu. Acetyl-coa and the regulation of metabolism: mechanisms and consequences. *Current opinion in cell biology*, 33:125–131, 2015.

48. Andrew Stuart and Anthony R Humphries. *Dynamical systems and numerical analysis*, volume 2. Cambridge University Press, 1998.

49. Rupali Vohra and Miriam Kolko. Neuroprotection of the inner retina: Müller cells and lactate. *Neural regeneration research*, 13(10):1741, 2018.
50. Eric J Warrant. Mammalian vision: Rods are a bargain. *Current Biology*, 19(2):R69–R71, 2009.
51. Barry S Winkler, Catherine A Starnes, Michael W Sauer, Zahra Firouzgan, and Shu-Chu Chen. Cultured retinal neuronal cells and müller cells both show net production of lactate. *Neurochemistry international*, 45(2-3):311–320, 2004.
52. Margaret Wong-Riley. Energy metabolism of the visual system. *Eye and brain*, 2:99, 2010.
53. Ying Yang, Saddek Mohand-Said, Aude Danan, Manuel Simonutti, Valérie Fontaine, Emmanuelle Clerin, Serge Picaud, Thierry Léveillard, and José-Alain Sahel. Functional cone rescue by rdcvf protein in a dominant model of retinitis pigmentosa. *Molecular Therapy*, 17(5):787–795, 2009.
54. Minfeng Ying, Cheng Guo, and Xun Hu. The quantitative relationship between isotopic and net contributions of lactate and glucose to the tricarboxylic acid (tca) cycle. *Journal of Biological Chemistry*, 294(24):9615–9630, 2019.
55. Kanako Yokosako, Tatsuya Mimura, Hideharu Funatsu, Hidetaka Noma, Mari Goto, Yuko Kamei, Aki Kondo, and Masao Matsubara. Glycolysis in patients with age-related macular degeneration. *The open ophthalmology journal*, 8:39, 2014.
56. Richard W Young. The renewal of photoreceptor cell outer segments. *The Journal of cell biology*, 33(1):61–72, 1967.
57. Richard W Young and Dean Bok. Participation of the retinal pigment epithelium in the rod outer segment renewal process. *The Journal of cell biology*, 42(2):392–403, 1969.

A Framework for Performing Data-Driven Modeling of Tumor Growth with Radiotherapy Treatment

Heyrim Cho, Allison L. Lewis, Kathleen M. Storey, Rachel Jennings, Blerta Shtylla, Angela M. Reynolds, and Helen M. Byrne

Abstract Recent technological advances make it possible to collect detailed information about tumors, and yet clinical assessments about treatment responses are typically based on sparse datasets. In this work, we propose a workflow for choosing an appropriate model, verifying parameter identifiability, and assessing the amount of data necessary to accurately calibrate model parameters. As a proof-of-concept, we compare tumor growth models of varying complexity in an effort to determine

Authors Heyrim Cho, Allison L. Lewis, and Kathleen M. Storey have equally contributed to this chapter.

H. Cho
University of California Riverside, Riverside, CA, USA
e-mail: heyrimc@ucr.edu

A. L. Lewis
Lafayette College, Easton, PA, USA
e-mail: lewisall@lafayette.edu

K. M. Storey
University of Michigan, Ann Arbor, MI, USA
e-mail: storeyk@umich.edu

R. Jennings
United Health Group, Minnetonka, MN, USA
e-mail: rljennings@ara.com

B. Shtylla
Department of Mathematics, Pomona College, Claremont, CA, USA
e-mail: blerta.shtylla@pomona.edu

A. M. Reynolds (✉)
Virginia Commonwealth University, Richmond, VA, USA
e-mail: areynolds2@vcu.edu

H. M. Byrne
University of Oxford, Oxford, UK
e-mail: helen.byrne@maths.ox.ac.uk

© The Association for Women in Mathematics and the Author(s) 2021   179
R. Segal et al. (eds.), *Using Mathematics to Understand Biological Complexity*,
Association for Women in Mathematics Series 22,
https://doi.org/10.1007/978-3-030-57129-0_8

the level of model complexity needed to accurately predict tumor growth dynamics and response to radiotherapy. We consider a simple, one-compartment ordinary differential equation model which tracks tumor volume and a two-compartment model that accounts for tumor volume and the fraction of necrotic cells contained within the tumor. We investigate the structural and practical identifiability of these models, and the impact of noise on identifiability. We also generate synthetic data from a more complex, spatially-resolved, cellular automaton model (CA) that simulates tumor growth and response to radiotherapy. We investigate the fit of the ODE models to tumor volume data generated by the CA in various parameter regimes, and we use sequential model calibration to determine how many data points are required to accurately infer model parameters. Our results suggest that if data on tumor volumes alone is provided, then a tumor with a large necrotic volume is the most challenging case to fit. However, supplementing data on total tumor volume with additional information on the necrotic volume enables the two compartment ODE model to perform significantly better than the one compartment model in terms of parameter convergence and predictive power.

Keywords Systems biology · Mathematical oncology · Parameter identifiability · Bayesian sequential calibration · Model selection

1 Introduction

Cancer remains one of the leading causes of death in the world, second only to cardiac disease. As such, it represents a significant global public health and socio-economic problem. Of particular interest, given the unpleasant side-effects that accompany many cancer treatments, is being able to establish as early as possible whether a patient will respond (or is responding) to a particular treatment and, based on this assessment, whether treatment should be continued or a new treatment started. Mathematical modeling provides a natural framework within which to answer such questions. In more detail, mechanistic models that describe the growth dynamics of a tumor and its response to treatment may be fit to patient data collected during treatment and used to predict how the tumor's size (and possibly composition) will change if treatment is continued. Model fits, parameter estimates, and predictions may be revised as treatment progresses and more patient data become available. These predictions can then be used to inform decisions about whether to continue with the current treatment.

The approach outlined above relies upon the availability of time-dependent mathematical models of tumor growth and patient data to which the models can be fit. Advances in technology mean that it is now possible to collect detailed information about tumors (e.g., their size, spatial composition, mutational status, vascularity and degree of immune infiltration). Even so, decisions about treatment options (e.g., surgery, radiotherapy, chemotherapy and immunotherapy), and assessments about treatment responses are often based on statistical analyses of sparse and noisy data relating to a small number of quantities of interest (e.g., tumor volumes at three time-points: at diagnosis, at treatment start, and at treatment end).

The availability of detailed information about tumors has undoubtedly stimulated the development of a large number of mathematical and computational models of tumor growth. These range from spatially-averaged, phenomenological models formulated using differential equations (e.g., logistic growth, Gompertzian growth) [8, 10], to multiphase models based on mixture theory [4, 20] or phase field theory [22], and multiscale models that couple subcellular, cellular, and tissue scale phenomena [14]. In addition to simulating tumor growth, these models have also been used to study tumor responses to treatments including radiotherapy, chemotherapy, immunotherapy and combinations thereof [3, 19]. Unfortunately, the absence of suitable experimental data coupled with the complexity of many of the theoretical models means that few of them have been validated and/or parameterized. Additionally, the sheer number and variety of available models makes it difficult to determine which model may be most appropriate for a given scenario and available data [9, 21, 24].

In this investigation, we propose a workflow for determining an appropriate model to be used under the constraints of a given scenario (i.e. when data is scarce or noisy). We begin with an identifiability analysis in two parts: structural identifiability, which determines whether the model parameters can be uniquely estimated in an "ideal world", and practical identifiability, which reassesses parameter identifiability in a real-world scenario, in which data is noisy and potentially sparse.

After establishing structural and practical identifiability, we tackle the question of which model to fit to the available data. We perform a parameter sweep to measure error in model predictions across different parameter regimes, establishing conditions under which certain models can be used to make accurate predictions about tumor growth. While we desire model simplicity whenever possible, we recognize that in some scenarios (for example, when the tumor comprises a large portion of necrotic tissue), a model that tracks only tumor volume (for instance) may not be able to describe the tumor growth dynamics.

Once we have chosen an appropriate model for a given scenario and verified that its parameters are structurally and practically identifiable, we examine how the model calibration is affected by the availability of data. In this investigation, we perform sequential calibration, adding one data point at a time to determine how much data is necessary to uniquely calibrate the model parameters. At each step, we use the current parameter values to predict the future tumor volume after treatment has concluded, and compare this prediction to our known "truth" to assess whether the current amount of data is adequate for making future predictions. This investigation helps us to determine the extent to which additional data will increase the predictive power of our mathematical models.

The approach outlined above serves as a "proof-of-concept" for our proposed framework; we test our procedure on two simple compartmental models for tumor growth: one that describes only total tumor volume, and a second that also accounts for the proportion of necrotic cells. We compare their dynamics to data generated from a more complex cellular automaton model (CA), which we use as our "truth". Additionally, we combine our growth models with the linear-quadratic model [12] to simulate tumour responses to a radiotherapy treatment regimen and also to test our model predictions in the presence of an intervention to tumor growth.

The remainder of this paper is structured as follows. In Sect. 2 we introduce the one- and two-compartment models that we use to predict tumor growth and response to radiotherapy. We also outline the structure of the cellular automaton model that we use to generate synthetic data for fitting the compartment models. We present typical CA simulation results that illustrate how a tumor's spatial composition and growth dynamics may change in response to radiotherapy. In Sect. 3, we explain how structural and practical identifiability analysis methods can be used to determine whether it is possible to infer the parameters associated with a particular model when the data is perfect (structural identifiability) and noisy (practical identifiability). We include several case studies that investigate how the addition of noise to the CA data affects the ability to recover model parameters, and compare the ability of the one- and two-compartment models to fit data generated by the CA under a variety of conditions, including a range of necrotic heterogeneity levels. In Sect. 4, we study the goodness of fit of the one-compartment model to synthetic data generated from the two-compartment model. Through an extensive search of parameter space, we determine how parameters in the one- and two-compartment models are related, and discuss how these results can be used to select an appropriate model that will yield accurate predictions while still maintaining model simplicity whenever possible. Finally, in Sect. 5 we perform sequential model calibration to determine how much data is needed to accurately infer model parameters. The paper concludes in Sect. 6 with a summary of our key results and directions for future work.

2   The Mathematical Models

Here we introduce the three mathematical models that will be used throughout this investigation. The first is a one-compartment ODE model that tracks tumor volume over time. It is the most basic model that we use to describe tumor growth. The second model is a two-compartment ODE model that incorporates a state variable for tracking the portion of tumor volume that is composed of necrotic tissue, thereby allowing for tumor heterogeneity. This model will be used both as a data generator to test the capabilities of the one-compartment ODE model as well as a model to be calibrated against "true" data. Our final and most complex model is an cellular automaton model (CA) that we use to generate our "truth" data, as it is assumed to most accurately reflect reality by including the cell cycle, quiescent cells, and oxygen levels [14]. In all three cases, we will incorporate treatment via radiation using the linear-quadratic model for radiotherapy.

2.1   The One-Compartment Model

Our one-compartment model describes the time evolution of the tumor volume, $V(t)$, using a logistic growth model with growth rate λ and carrying capacity K:

$$\frac{dV}{dt} = \underbrace{\lambda V\left(1 - \frac{V}{K}\right)}_{\text{logistic growth}} - \underbrace{\eta V.}_{\text{natural cell death}} \tag{1}$$

Natural cell death, at rate η, is incorporated via the term $-\eta V$. In what follows, it will be convenient to re-parameterize Equation (1) to obtain the simpler form

$$\frac{dV}{dt} = AV\left(1 - \frac{B}{A}V\right), \tag{2}$$

where $A = \lambda - \eta$ and $B = \frac{\lambda}{K}$. From here on, we will refer to the one-compartment model in its re-scaled form, in Equation (2).

This simple model views the tumor as a homogeneous mass in which all cells are identical. In practice, however, as the tumor grows, regions at a distance from oxygen and nutrient sources (e.g., blood vessels for tumors growing in vivo) may undergo necrotic cell death in response to sustained oxygen and/or nutrient deprivation. In the one-compartment model, such dead or necrotic cells are assumed to be removed from the tumor instantaneously.

2.2 The Two-Compartment Model

In order to account for some aspects of tumor heterogeneity, we study a two-compartment model that tracks the time evolution of the volume of viable tumor cells ($V(t)$) and the volume of the necrotic core ($N(t)$), and that was originally developed in [18]. The population of proliferating (i.e., viable) cells is assumed to grow logistically with growth rate λ and carrying capacity K. Additionally, we assume that viable cells convert to necrotic cells at a constant rate η, and that necrotic material undergoes natural decay at a constant rate ζ. Combining these processes, we arrive at the following ODE system for $V(t)$ and $N(t)$:

$$\frac{dV}{dt} = \lambda V\left(1 - \frac{V}{K}\right) - \eta V, \tag{3a}$$

$$\frac{dN}{dt} = \eta V - \zeta N. \tag{3b}$$

To facilitate comparison with the one-compartment model, we reformulate (3a)–(3b) in terms of $Y(t)$, the total tumor volume ($Y = V + N$), and $\Phi(t)$, the fraction of the total volume that comprises necrotic cells ($\Phi = N/Y$). Using this notation, Equations (3a) and (3b) can be rewritten in the form

$$\frac{dY}{dt} = \lambda(1 - \Phi)Y\left(1 - (1 - \Phi)\frac{Y}{K}\right) - \zeta\Phi Y, \tag{4a}$$

$$\frac{d\Phi}{dt} = (1 - \Phi)\left[\eta - \lambda\Phi\left(1 - (1 - \Phi)\frac{Y}{K}\right) - \zeta\Phi\right]. \tag{4b}$$

We note that in the limit as $\zeta \to \infty$ (i.e., if the timescale for degradation of necrotic material is much shorter than the timescale for other processes included in the model), Equations (4a) and (4b) reduce to the one-compartment model defined by Equation (1).

2.3   The Cellular Automaton Model

We use a spatially-explicit, hybrid cellular automaton model (CA) to generate synthetic tumor volume data. Our cellular automaton model is adapted from the one developed in [14]. In the CA model, cells are arranged on a 200×200 grid, which represents a two-dimensional cross-section of size 0.36×0.36 cm^2 through a three-dimensional spheroid in vitro. Each automaton can be occupied either by a tumor cell or culture medium. The CA couples the dynamics of automaton elements arranged on the grid to the oxygen concentration. We identify with each automaton $\mathbf{x} = (x, y)$ at time t a dynamical variable with a state and a neighborhood. The four possible states are proliferating $\mathscr{P}$, quiescent $\mathscr{Q}$, necrotic $\mathscr{N}$, and empty $\mathscr{E}$, determined by the oxygen level at each site: if $c > c_Q$ then the cells proliferate, if $c_N < c < c_Q$ then the cells stop proliferating and halve the rate at which they consume oxygen, and if $c \leq c_N$ then the cells become necrotic (see Table 1 for details). Each cell communicates with cells within its Moore neighborhood, i.e., its eight nearest neighbors. Proliferating cells are assigned counters that describe

Table 1 A summary of the parameters used in the CA and their default values. Parameter values are estimated using experimental data from the prostate cancer cell line, PC3, in [14]

Parameter	Description	Value	Units
l	Cell size	0.0018	cm
L	Domain length	0.36	cm
$\bar{\tau}_{cycle}(\sigma_{cycle})$	Mean (standard deviation) cell cycle time	18.3 (1.4)	h
c_∞	Background O$_2$ concentration	2.8×10^{-7}	mol cm^{-3}
D	O$_2$ diffusion constant	1.8×10^{-5}	cm^{2}s^{-1}
c_Q	O$_2$ concentration threshold for proliferating cells	1.82×10^{-7}	mol cm^{-3}
c_N	O$_2$ concentration threshold for quiescent cells	1.68×10^{-7}	mol cm^{-3}
κ_P	O$_2$ consumption rate of proliferating cells	1.0×10^{-8}	mol cm^{-3}s^{-1}
κ_Q	O$_2$ consumption rate of quiescent cells	5.0×10^{-9}	mol cm^{-3}s^{-1}
p_{NR}	Rate of lysis of necrotic cells	Range: 0.004–0.016	hr^{-1}

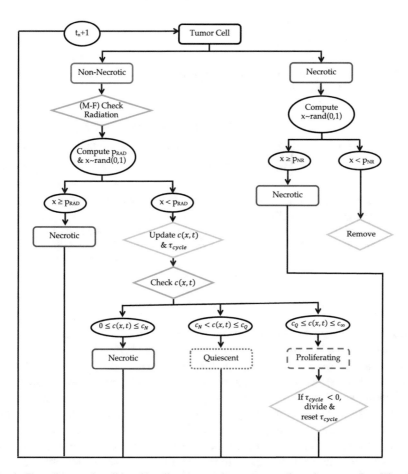

Fig. 1 Flow diagram describing the cell movement between necrotic, quiescent and proliferating states. The parameter p_{RAD} denotes the probability that a cell becomes necrotic following radiotherapy and $p_{RAD} = 1 - e^{-\alpha d - \beta d^2}$. The parameter τ_{cycle} denotes the specific cell's assigned cycle length

where they are in the cell cycle. These counters are initially drawn from a normal distribution with mean $\bar{\tau}_{cycle}$ and standard deviation σ_{cycle}. After each time step, the cell cycle counter of each cell decreases by an amount that depends on the number of neighboring cells; a smaller reduction in cell cycle time occurs with a larger number of neighbors, to model the regulatory process known as contact inhibition of proliferation. Figure 1 summarizes how a cell can transition between quiescent, proliferating and necrotic states.

We model the single growth-rate-limiting nutrient, oxygen, explicitly via a reaction-diffusion equation. In particular, the evolution of the oxygen concentration $c(\mathbf{x}, t)$ (mol cm^{-3}) at location $\mathbf{x}$ for time t is described by:

$$\frac{\partial c(\mathbf{x}, t)}{\partial t} = D\nabla^2 c(\mathbf{x}, t) - \Gamma(\mathbf{x}, t), \tag{5}$$

where D is the oxygen diffusion coefficient (cm^2 s^{-1}). and $\Gamma(\mathbf{x}, t)$ is the oxygen consumption rate (mol cm^{-3} s^{-1}), defined as follows:

$$\Gamma(\mathbf{x}, t) = \begin{cases} \kappa_P & \text{if } \mathbf{x} \text{ is occupied by a proliferating cell} \\ \kappa_Q & \text{if } \mathbf{x} \text{ is occupied by a quiescent cell} \\ 0 & \text{otherwise.} \end{cases}$$

The parameters κ_P and κ_Q are the rates at which quiescent and proliferating cells consume oxygen, respectively, and are in mol cm^{-3} s^{-1} with $\kappa_P >= \kappa_Q$. We also use the following initial and boundary conditions to describe the situation in which oxygen diffuses from the boundaries of a square Petri dish into the culture medium:

$$c(x, y, 0) = c_\infty,$$
$$c(0, y, t) = c(L, y, t) = c(x, 0, t) = c(x, L, t) = c_\infty,$$

where L is the domain length, and c_∞ is the background O$_2$ concentration. See Table 1 for a list of the parameter values used to simulate the CA.

When a cell cycle counter reaches 0, the cell divides to produce two identical cells, one located at the same site as the parent, and one placed in an empty neighboring site, if available. If more than one neighboring site is empty, the site with the maximum number of neighbors is chosen to maintain cell-cell adhesion. If no adjacent sites are empty, then in an effort to simulate the mechanical stress exerted on neighboring cells during spheroid expansion, we find the shortest chain of cells connecting the dividing cell to the spheroid's boundary and shift this chain outward to create space for the daughter cell. Figure 2a, b displays this process.

A cell becomes necrotic if the oxygen concentration at its location falls below a specified threshold, or if the cell is irradiated, as discussed in Sect. 2.4. Necrotic cells are lysed at rate p_{NR}. Lysis involves removing the necrotic cell and then shifting inward a chain of cells starting from the boundary of the spheroid to fill in the removed cell's site. The spheroidal shape of the tumor is preserved by choosing the boundary cell that is farthest from the spheroid center, and then shifting a chain of adjacent cells inward. Figure 2c, d displays this process.

All model simulations are initialized by placing a circular cluster of cells in the center of the grid: this imitates seeding a spheroid in a Petri dish. The cells consume oxygen as it diffuses from the culture medium and this enables them to progress through the cell cycle and to divide. As the spheroid grows, oxygen levels at its center fall. When the oxygen concentration drops below a threshold value cells exit the cell cycle and become quiescent. As the spheroid grows further, and oxygen levels decrease further, quiescent cells die via necrosis and the resulting necrotic debris is transported away from the spheroid.

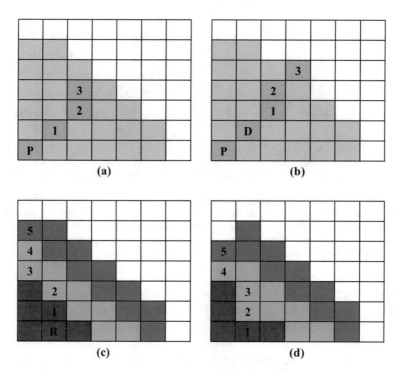

(a)

(b)

(c)

(d)

Fig. 2 Figures adapted from [14]. The spheroidal shape of the tumor is preserved by shifting a chain of cells outward when cell division occurs, shown (**a**) before and (**b**) after division, and by pulling a chain of cells inward when a cell lyses, shown (**c**) before and (**d**) after lysis. In (**a**)–(**b**), P denotes the dividing parent cell, and D denotes the daughter cell. The dividing cell pushes the chain of cells in (**a**), labeled by 1,2,3, outward to occupy the sites shown in (**b**). In (**c**)–(**d**), R denotes the cell that is removed. The numbered chain of cells, 1–5, are shifted in order to take the place of cell R

We use the CA to generate a series of synthetic spheroids which differ in their growth rates, sizes and spatial composition. Parameters are set to baseline values determined using experimental data from the prostate cancer cell line, PC3, in [14]. The results presented in Fig. 3 show how the size and composition of a spheroid change over 60 days of growth as we vary p_{NR}, the rate at which necrotic cells are removed. We use $p_{NR} = 0.015\,\text{hr}^{-1}$ and $p_{NR} = 0.004\,\text{hr}^{-1}$ to generate control (untreated) spheroids with approximately 20% and 50% necrotic volume, respectively, at steady state.

2.4 Radiotherapy Treatment

We now explain how we incorporate treatment with radiotherapy (RT) in all three models. We consider a typical tumor treatment regimen in which daily doses of 2 Gy

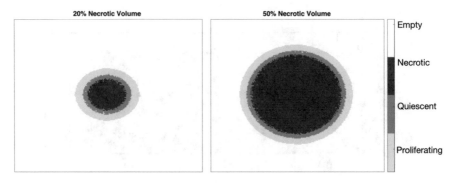

Fig. 3 CA simulations after 60 days in the absence of treatment, representing two-dimensional cross-sections of tumor spheroids. On the left, we used $p_{NR} = 0.015$ to generate a spheroid in which the necrotic cells occupy approximately 20% of total tumor volume. On the right, we used $p_{NR} = 0.004$ to generate a spheroid in which necrotic cells occupy approximately 50% of total tumor volume

are administered Monday through Friday for 6 weeks. We use the linear-quadratic model [12] to account for the effect of RT. This model assumes that the fraction of cells that survive exposure to a dose d of RT is given by

$$\text{Survival fraction,} \quad SF = e^{-\alpha d - \beta d^2}, \tag{6}$$

where α and β are tissue specific radiosensitivity parameters that model single and double strand breaks of the DNA [17]. We assume that the effect of RT is instantaneous, with the non-surviving cell fraction immediately removed when therapy is administered. Under these assumptions, the one-compartment model becomes

$$\frac{dV}{dt} = AV\left(1 - \frac{B}{A}V\right) \quad \text{for } t_i^+ < t < t_{i+1}^-, \tag{7}$$

$$V(t_i^+) = \exp(-\alpha d - \beta d^2)\, V(t_i^-).$$

where t_i (for $i = 1, 2, \ldots, N$) denote the times at which radiotherapy is delivered, and $V(t_i^{\pm})$ denote the tumour volume just before and after radiotherapy is administered.

Treatment in the two-compartment model is modeled analogously, except that the sink of irradiated cells from the viable tumor volume will have an equal and opposite source term in the ODE for the necrotic component. Similarly in the cellular automaton model, each living cell becomes necrotic with probability $1 - e^{-\alpha d - \beta d^2}$ when radiotherapy is administered.

2.5 Typical CA Simulation Results

We generate synthetic spheroids using the CA described in Sect. 2.3, using the default parameter values listed in Table 1. Snapshots of typical CA tumor spheroids in the absence of treatment at days 6, 17, 30 and 70 are presented in Fig. 4a. Similar results showing the response to radiotherapy of tumors with low radiosensitivity are presented in Fig. 4b and for tumors with high radiosensitivity in Fig. 4c. Tumors with faster necrotic decay ($p_{NR} = 0.015$) are shown in the four-panels on the left, and tumors with slower necrotic decay ($p_{NR} = 0.004$) are shown on the right. We note that when radiotherapy is applied, treatment begins on Day 15. We observe that, in most cases, the size of the tumor on Day 70 is larger when the rate of necrotic decay is low than when it is high. However, if the tumor has high radiosensitivity, then the situation reverses: all living cells are eliminated when the necrotic cells decay slowly but not when they decay rapidly.

For each set of parameter values, we simulate 100 realizations of the CA and determine how the mean total tumor volume and mean necrotic volume change over time. Figure 5a presents the averaged results when no treatment is applied, with results corresponding to the high rate of necrotic decay ($p_{NR} = 0.015$) on the left, and results corresponding to the low rate ($p_{NR} = 0.004$) on the right. Figures 5b, c summarize the results for both cases when radiotherapy is applied. In Fig. 5b, the tumor has a lower radiosensitivity level ($\alpha/\beta = 9$), while Fig. 5c shows the results for tumors with higher radiosensitivity ($\alpha/\beta = 1$). The plots of mean tumor volume and mean necrotic volume confirm that the trends we observe in a single realization (see Fig. 4a–c) are representative of the average behavior in each case. We use the synthetic data from these representative simulations to calibrate the ODE models in Sects. 3, 4, and 5.

3 Identifiability Analysis

3.1 Structural Identifiability Analysis

The concept of structural identifiability was first introduced in 1970 by Bellman and Astrom [2]; they asked whether, given perfect input data and a measured output signal that relates to available experimental data, it is possible to determine parameters associated with a dynamical systems model. A model identification question then asks whether it is possible to uniquely recover all model parameters given sufficient, error-free data about one (or more) model outputs. We note here that this type of analysis is sometimes referred to as structural identifiability analysis, as it relies solely on the properties of the dynamical system and respective model observable outputs. It should not be confused with practical identifiability analysis, which is concerned with the ability to recover parameter values from error-prone experimental data and depends on the computational approach used to parameterize

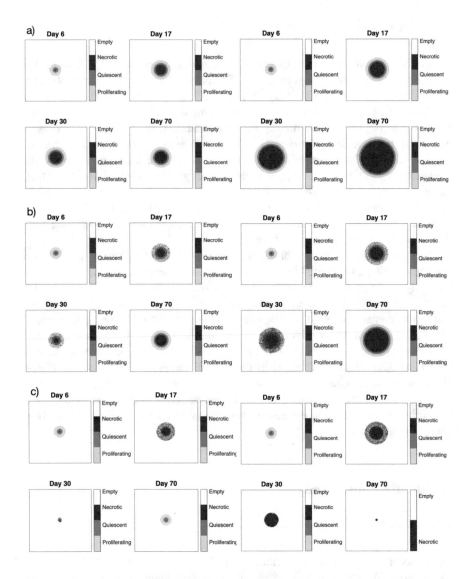

Fig. 4 (**a**) *No treatment.* Tumor composition on Days 6, 17, 30, and 60 when the rate of necrotic decay is high (left) and low (right), and no treatment is applied. (**b**) *Low radiosensitivity* ($\alpha/\beta = 9$). Tumor composition on Days 6, 17, 30, and 60 when the rate of necrotic decay is high (left) and low (right), and radiotherapy is applied. (**c**) *High radiosensitivity* ($\alpha/\beta = 1$). Tumor composition on Days 6, 17, 30, and 60 when the rate of necrotic decay is high (left) and low (right), and radiotherapy is applied

the model. It is important to note that many difficulties related to estimating parameters by fitting mathematical models to datasets may stem from lack of structural identifiability. In such cases, the system may not admit unique parameter

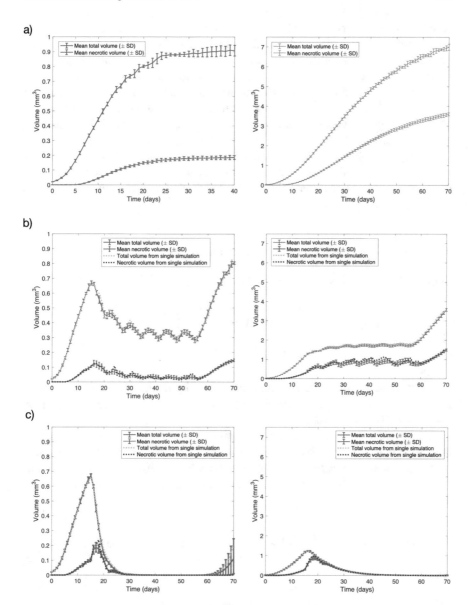

Fig. 5 (**a**) *Mean CA spheroid volume with no treatment.* The tumor grows in the absence of radiotherapy, when the rate of necrotic cell death is high (left), and low (right). (**b**) *Mean CA spheroid volume with low radiosensitivity.* The tumor exhibits radiosensitivity level $\alpha/\beta = 9$, when the rate of necrotic cell death is high (left), and low (right). (**c**) *Mean CA spheroid volume for tumor cells with high radiosensitivity.* The tumor exhibits radiosensitivity level $\alpha/\beta = 1$, when the rate of necrotic cell death is high (left), and low (right). The plots in Fig. 5 show the time evolution of the mean total tumor volume and mean necrotic volume, obtained by averaging over 100 realizations of the CA. The error bars indicate the standard deviation at each time point. The dashed lines show the total tumor volume and the necrotic volumes from the representative simulation used to fit to the ODE models. The figures on the left correspond to spheroids with high rates of necrotic decay ($p_{NR} = 0.015$) and small necrotic volumes; those on the right correspond to spheroids with low rates of necrotic decay ($p_{NR} = 0.004$) and large necrotic volumes

sets for a given observed model output. For further details about the techniques we use, we refer the interested reader to [5, 6]. In the remainder of this section we study the structural identifiability of the one- and two-compartment models presented in Sect. 2. Due to the form of the treatment terms, we employ the Taylor series approach [5, 23] to establish model structural identifiability and then compare our results with those obtained through a generating series approach implemented in Matlab through the GenSSI package [7]. To the best of our knowledge, assessing and characterising the structural identifiability of CAs remains an open problem. One approach would be to derive mean field descriptions of the CAs and to perform identifiability analysis of the resulting equations. Alternatively, several authors have shown how Approximate Bayesian Computation can be used for parameter inference of CAs [15, 16].

3.1.1 One-Compartment Model: No Radiation

In this case the model reduces to logistic growth, with

$$\frac{dV}{dt} = AV\left(1 - \frac{B}{A}V\right),$$ (8)

$$V(0) = V_0.$$

We wish to establish the structural identifiability of the model with unknown parameters $p = \{A, B\}$, observable quantity $y(t; p) = V(t)$ and known initial conditions $V(t = 0) = V_0$.

Before we state results, we briefly outline the Taylor series approach, as delineated in [5, 23]. We assume that the observation function $y(t; p)$ is analytic in a neighborhood of some time. Then we can evaluate $y(t; p)$ and its successive time derivatives in terms of the model parameters and initial conditions at time $t = 0^+$ using

$$y(t, p) = y(0^+; p) + y^{(1)}(0^+; p)t + y^{(2)}(0^+; p)\frac{t^2}{2!} + \ldots + y^{(i)}(0^+; p)\frac{t^i}{i!} + \ldots,$$ (9)

where

$$y^{(i)}(0^+; p) = \frac{d^i y}{dt^i}(0^+; p).$$ (10)

Given Equations (9) and (10), the problem reduces to identifying a system of algebraic equations that relate the unknown model parameters (here A and B) to known values of the observable $y(t; p)$ and its derivatives at $t = 0^+$. If these equations admit unique solutions for each parameter, we consider the system to be structurally identifiable; otherwise, it is either locally identifiable or non-identifiable.

Since for our problem only A and B are unknown, we start by computing the first two Taylor series coefficients:

$$V'(0^+) = AV(0^+)\left(1 - \frac{B}{A}V(0^+)\right) \tag{11}$$

$$V''(0^+) = AV'(0^+) - 2BV(0^+)V'(0^+). \tag{12}$$

Denoting by $a_0 = V(0^+)$, $a_1 = V'(0^+)$ and $a_2 = V''(0^+)$, we obtain the following pair of simultaneous equations for the unknown parameters A and B:

$$a_1 = Aa_0 - Ba_0^2 \tag{13}$$

$$a_2 = a_1 A - 2a_1 Ba_0, \tag{14}$$

with solution

$$A = 2\frac{a_1}{a_0} - \frac{a_2}{a_1} \quad \text{and} \quad B = \frac{a_1}{a_0^2} - \frac{a_2}{a_1 a_0}. \tag{15}$$

Since a_0, a_1, and a_2 are known, the variables A and B are globally structurally identifiable except for at most a set of points of zero measure (i.e., points for which $a_0 = 0$, $a_1 = 0$). We remark that in this case we have exploited the reduced number of parameters due to model rescaling. For the original, dimensional model (1), the three parameters $p = \{\lambda, K, \eta\}$ are not uniquely identifiable: the Taylor series coefficients do not contain enough information to uniquely extract the three parameters from tumor volume observations.

Next, we repeat the above analysis using the generating series approach implemented in GenSSI [7]. Briefly, GenSSI implements a generating series approach coupled with identifiability tableaus [1] for linear and non-linear systems of ODEs. The underlying principle is to obtain equations for model parameters by computing successive Lie derivatives of the right hand side of the ODE system and model observable quantities ($y(t; p) = V(t)$ in our case). If the solution of the system of parameter equations is unique then the parameters are declared globally identifiable. We implemented our model in GenSSI and confirmed that $\{A, B\}$ in (2) are structurally identifiable, whereas $\{\lambda, K, \eta\}$ in (1) are not structurally identifiable. These results are consistent with results of the Taylor series approach outlined above.

3.1.2 One-Compartment Model: Point Radiation Treatment

The problem now reads

$$\frac{dV}{dt} = AV\left(1 - \frac{B}{A}V\right) \quad \text{for} \quad t_i^+ < t < t_{i+1}^-, \tag{16}$$

$$V(t_i^+) = \exp(-\alpha d - \beta d^2)\, V(t_i^-).$$

Since, in this case, the experimental observable $y(t; p) = V(t)$ corresponds to tumor volume, parameters A and B are identifiable from observations of the system without therapy (see Equation (15)). In addition, for a radiotherapy treatment the timing and dose t_i and d are also known quantities. Thus, the unknown parameters to identify from the tumor volume measurements are α and β. In order to identify α and β, we should compute, in principle at least, two Taylor series coefficients by employing a Taylor series expansion around the treatment time, t_i (note that we use one-sided limits and derivatives at $t_i^{\pm}$, as in the previous section).

Using Equation (16), it is straightforward to show:

$$A_1 = AA_0\Gamma - BA_0^2\Gamma^2, \tag{17}$$

$$A_2 = A_1(A - 2BA_0\Gamma), \tag{18}$$

where $A_0 = V(t_i^-)$, $A_1 = V'(t_i^+)$, and $A_2 = V''(t_i^+)$ are known, and $\Gamma = \exp(-\alpha d - \beta d^2)$.

Inspection of Equations (17) and (18) reveals that they do not admit unique solutions for α and β since α and β appear in both equations via the parameter grouping Γ. We declare α and β to be non-identifiable in this setting (the same results were obtained using GenSSI implementation). Therefore, in what follows, we fix α and vary β. This is reasonable since radiosensitivity of cancer is often characterized by the ratio α/β; we vary β to allow α/β to take on a range of values. In practice, estimates of the values of α and the ratio α/β could be obtained, for a particular biological tissue, by measuring the volume reduction caused by exposure to different radiotherapy doses and fitting these data to the linear quadratic model.

3.1.3   Two-Compartment Model

The two compartment model with treatment is given by the following model equations:

$$\frac{dY}{dt} = \lambda(1 - \Phi)Y\left(1 - (1 - \Phi)\frac{Y}{K}\right) - \xi\Phi Y \tag{19}$$

$$\frac{d\Phi}{dt} = (1 - \Phi)\left[\eta - \lambda\Phi\left(1 - (1 - \Phi)\frac{Y}{K}\right) - \xi\Phi\right], \quad \text{for} \quad t_i^+ < t < t_{i+1}^- \tag{20}$$

$$\Phi(t_i^+) = \Phi(t_i^-) + (1 - \Phi(t_i^-))(1 - \exp(-\alpha d - \beta d^2)). \tag{21}$$

Since the calculations are similar to, but more involved than those used for the one-compartment model, the details are presented in the Appendix. For completeness, we summarize our findings here. In the absence of treatment, with

$p = \{\lambda, K, \xi, \eta\}$, observable quantities $y(t; p) = \{Y(t), \Phi(t)\}$, and known initial conditions, we obtain unique solutions of the unknown parameters in terms of the observable quantities and their derivatives; we thus declare all four parameters structurally identifiable. We also repeat the analysis in the case in which only the tumor volume is observed, i.e., $y(t; p) = Y(t)$, but with known initial conditions for the tumor volume and necrotic fraction. In this case, we take higher order Taylor series coefficients (up to order 4) and find that $p = \{\lambda, K, \xi, \eta\}$ are structurally identifiable. GenSSI calculations confirm our findings.

When treatment is added, as detailed in the Appendix, we find that the radiation parameters α and β are not structurally identifiable, in agreement with the analysis for the single compartment model.

3.2 Practical Identifiability Analysis and Parameter Estimation

After establishing structural identifiability, it is natural to consider the practical identifiability of a model's parameters. In particular, given experimental data with measurement noise and a specific model, is it possible to uniquely determine a set of model parameter values that are most likely to produce the data?

One approach for determining the practical identifiability of the parameters while performing model calibration is through the use of a Metropolis algorithm, based on Markov Chain Monte Carlo (MCMC) techniques. Here, we construct Markov chains whose stationary distributions coincide with the posterior density of the parameters; thus, by sampling realizations of our parameters from this chain, we are effectively sampling from the parameter posterior density. The traditional Metropolis algorithm, as outlined in [25], constructs the posterior chains by drawing the next candidate, q^*, from a proposal function $J(q^*|q^{k-1})$, where q^{k-1} represents the previous parameter candidate. The goal of the Metropolis algorithm is to identify the set of parameter values that maximizes the likelihood function. This is equivalent to minimizing the sum-of-squares of the differences between the observed and predicted data. If the new candidate yields a smaller sum-of-squares error than the previous one, it is accepted as the next value in the posterior chain. Otherwise, we reject with some specified probability—see [25] for details—and the new state is taken to be the same as the old one, $q^k = q^{k-1}$. The traditional Metropolis algorithm assumes a symmetric proposal function J with respect to each of the individual parameters, though the Metropolis-Hastings algorithm [25] allows for asymmetric proposal functions. Here, we use the symmetric proposal function $J(q^*|q^{k-1}) = \mathcal{N}(q^{k-1}, C)$, where C is the covariance matrix for the parameter set.

In what follows, we perform model calibration using the Delayed Rejection Adaptive Metropolis (DRAM) algorithm for parameter estimation [11]. This extension of the traditional Metropolis algorithm includes two additional steps. The first, the delayed rejection step, allows for the proposal of an alternative parameter candidate from a narrower proposal distribution in place of outright rejection of the original candidate. This results in greater mixing in our posterior MCMC chains by preventing the algorithm from stagnating on a single accepted candidate for long periods of time while multiple new candidates are rejected. During the adaptation step, a periodic adaptation of the parameter covariance matrix is performed to incorporate new information gained from accepted candidates. This covariance matrix is built into our proposal distribution; recall, we use $J(q^*|q^{k-1}) = \mathcal{N}(q^{k-1}, C)$, where C is the parameter covariance matrix. Thus, periodically updating C to reflect new information about the accepted parameter candidates will result in quicker convergence to the posterior densities. In this investigation, we use an adaptation interval of every 100 parameter candidates. For further information about the DRAM algorithm, we refer the interested reader to [11, 25].

After completing the parameter estimation process, we look for evidence of successful parameter recovery. First, we investigate the MCMC posterior chains for good mixing; we desire posterior chains that resemble white noise to suggest that the entire parameter space has been explored without extended stagnation on certain candidates. For visual examples of well-mixed posterior chains, we refer the reader to [25]. We also consider the pairwise parameter plots, as these can illustrate identifiability issues in several ways. Pairs of parameters whose chains are highly correlated in a strictly linear fashion are said to be unidentifiable in the sense that they cannot be uniquely identified by calibration with the available data; infinitely many pairs of parameter values would yield the same model response. Identifiability issues can also manifest as posterior densities that are unchanged from the specified prior distributions, indicating that the parameters are uninformed by the available data. By considering the above indicators, we can determine whether the quality and quantity of the data is sufficient to support the unique identification of all model parameters.

3.2.1 The Impact of Necrotic Fraction on Model Calibration

We now investigate the ability of the one- and two-compartment ODE models to fit synthetic data generated from the CA for different values of the CA parameter p_{NR}, the probability of removal of a necrotic cell. We begin by generating synthetic data in the form of a single representative realization of the CA with $\alpha/\beta = 1$ and all other parameters fixed at the nominal values provided in Table 1. For calibration, we use data from the first day of weeks 2–7 and day 70 (i.e. days 8, 15, 22, 29, 36, 43, 50, and 70), corresponding to the first day of treatment each week and then a post-treatment scan to check for tumor regrowth. We feed this data to the relevant ODE model and estimate all parameters (A, B, and β for the one-compartment ODE model; λ, k, η, ζ, and β for the two-compartment model). When investigating the performance of the two-compartment ODE model, we consider two cases: (1) providing tumor volume and necrotic fraction data, and (2) providing only tumor volume data, but still estimating necrotic fraction in the absence of that data.

The results presented in Fig. 6a show that the one-compartment ODE fit improves as p_{NR} increases, suggesting that the one-compartment ODE is better able to model

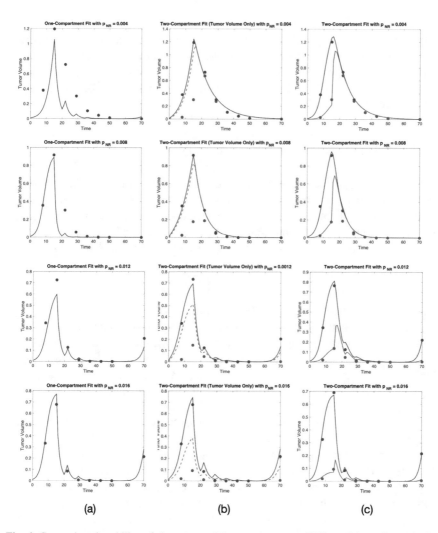

Fig. 6 Comparing the ability of the one- and two-compartment ODE models to fit synthetic data collected from the CA for varying values of p_{NR}, 0.004, 0.008, 0.012, and 0.016. (**a**) One-compartment model fit to tumor volume data from the CA model. (**b**) Two-compartment model fit to CA tumor volume data without necrotic core fraction data. (**c**) Two-compartment model fit to the CA data (for both tumor volume and necrotic fraction). Blue represents tumor volume; red represents necrotic fraction. Solid curves are fit to the given data; dashed curves show necrotic fraction estimate from the two-compartment ODE model generated in absence of necrotic data

scenarios in which the necrotic portion of the tumor is minimal. In contrast, the two-compartment ODE with both sets of data supplied fits the data with reasonable accuracy in all four cases, regardless of the p_{NR} value—see Fig. 6c. When the necrotic fraction data is not supplied, the two-compartment model generates good fits to total tumor volume data alone, for all values of p_{NR}. However, in all cases it vastly overestimates the necrotic fraction towards the beginning of the observation period, as shown in Fig. 6b. Thus, using only tumor volume data, we can reliably use the two-compartment model to make predictions about tumor volume, but cannot rely on the calibration to produce an accurate portrayal of the tumor heterogeneity. Parameter values generated by fitting the models to the CA data for various values of p_{NR} are given in Table 2.

3.2.2   The Impact of Noise on Parameter Recovery

In this case study, we seek to understand how the addition of noise to the data affects our ability to recover model parameters; specifically, at what point does the noise level overcome our ability to determine what parameter values were used to generate the data? We generate noise uniformly on an interval centered over each data point, with a range depending on the level of noise desired; that is, at each time t,

$$y_{\text{noise}}(t) = y_{\text{exact}}(t)(1 + \varepsilon),$$

where $\varepsilon \sim \mathscr{U}(-x, x)$ if we desire $100x\%$ noise.

We generate noisy synthetic data for the one-compartment and two-compartment models in the presence and absence of radiotherapy, and then calibrate the data against the model used for its generation, varying the noise level from 1% to 20%. The results are presented in Fig. 7, separated into three cases for easier visualization: (a) one-compartment model, (b) two-compartment model with a low necrotic fraction (approximately 20% necrotic tissue over the long term without treatment), and (c) two-compartment model with a high necrotic fraction (approximately 50% necrotic tissue over the long term without treatment). For the one-compartment model, we estimate the parameter set $[A, B, \beta]$ and compare to the true parameter set $[0.5, 2, \beta]$, where the value of β used to generate the data depends on the radiosensitivity level specified. For the two-compartment model, we estimate $[\lambda, K, \eta, \zeta, \beta]$ and compare to the true parameter set $[1, 0.5, 0.5, 0.5, \beta]$ in the high-necrotic case, and $[1, 0.5, 0.5, 2, \beta]$ in the low-necrotic case. In all cases, we fix $\alpha = 0.14$ and estimate the radiosensitivity ratio α/β by varying β only, since we encounter structural identifiability issues when trying to fit both α and β simultaneously—see Sect. 3.1. Since we are interested in observing the behavior for a variety of radiosensitivity levels, we consider four cases for each model: no-treatment, high radiosensitivity ($\alpha/\beta = 1$, so $\beta = 0.14$), medium radiosensitivity ($\alpha/\beta = 3$, so $\beta = 0.0467$), and low radiosensitivity ($\alpha/\beta = 9$, so $\beta = 0.0156$). Data is supplied for calibration for two pre-treatment times (days 8 and 15), five treatment times (during days 22–50), and one post-treatment time (day 70). In each case, we measure the average relative error in parameter estimates (we compare the parameter values used to generate the data to those obtained via the parameter estimation procedure) over 10 runs, for four levels of noise: 1%, 5%, 10%, and 20%. As seen in Table 3 and Fig. 7, in all cases there is a positive correlation between the

Table 2 Parameter values for fits of one- and two-compartment models (with and without necrotic fraction data included) to the CA for varying p_{NR} values. These are the values of the parameters used to generate the graphs in Fig. 6

	One-comp. model				Two-comp. w/o Nec. data				Two-comp. w/o Nec. data			
p_{NR}	0.004	0.008	0.012	0.016	0.004	0.008	0.012	0.016	0.004	0.008	0.012	0.016
A	0.339	0.513	0.451	0.528								
B	0.111	0.552	0.703	0.668								
λ					1.696	2.724	2.959	2.749	0.565	0.517	0.570	0.691
k					2.975	2.639	1.358	2.572	1.138	1.106	0.673	0.758
η					1.459	2.529	2.542	2.340	0.074	0.073	0.133	0.136
ζ					0.098	0.171	0.898	2.177	0.124	0.202	0.639	1.144
β	0.113	0.243	0.140	0.170	0.470	0.426	0.123	0.116	0.235	0.245	0.132	0.185

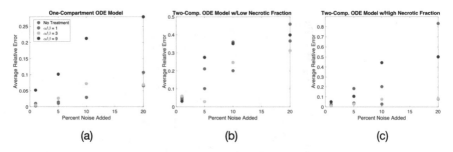

Fig. 7 Average relative error between the parameter values used to generate the synthetic data and those obtained by fitting to noisy data, averaged over 10 iterations in each trial for four different noise levels. (**a**) One-compartment ODE model with varying noise. (**b**) Two-compartment ODE model with low necrotic fraction ($\zeta = 2$) and varying noise. (**c**) Two-compartment ODE model with high necrotic fraction ($\zeta = 0.5$) and varying noise. In general, increased noise levels lead to higher relative errors

noise level and the average relative error in the parameter estimates. That is, as the level of noise in the data increases, the accuracy with which the true parameter values can be recovered decreases.

In Fig. 8, we focus on the one-compartment model with treatment and a low α/β ratio. For each noise level, we plot the posterior densities of our three parameters post-calibration. With 1% noise, the posterior distributions are centered at the values used to generate the data and are extremely well-informed, as illustrated by their narrow posterior densities. As the noise level increases, the posterior distributions widen, indicating less well-informed parameter estimates with greater variability. In addition, the distributions tend to drift rightwards as the noise level increases, suggesting that all parameter values are being overestimated. While we cannot be sure as to the cause of this "drifting" effect, we hypothesize that it may be due to our defining of the noise on a local scale, such that pre-treatment data will be "noisier" on average than data later on in the observation period. Further work is required to confirm this causation; meanwhile, we should remain cautious when dealing with noisy pre-treatment data, as it may play a significant role in the calibration of the parameters.

4 Model Selection

We now move to the question of determining how to choose the model most appropriate for the available data. When patient data is collected, we seek to identify a model that can be well calibrated to the data and make accurate predictions, while also retaining simplicity in terms of the underlying mechanisms and number of parameters. Minimizing the number of variables and parameters is desirable, due to the cost of collecting the data and conducting parameter estimation. For example,

Table 3 Mean relative error in parameter estimates, averaged over 10 runs in each category. High, Med and Low categories indicate radiosensitivity, corresponding to $\alpha/\beta = 1$, $\alpha/\beta = 3$, $\alpha/\beta = 9$, respectively. As the noise level increases, the average relative error tends to increase as well, indicating that parameter recovery is less successful

	Average relative error in parameter estimates											
	One-comp. model				Two-comp. w/o Low Nec.				Two-comp. w/o High Nec.			
Noise	None	High	Med	Low	None	High	Med	Low	None	High	Med	Low
1%	0.003	0.010	0.005	0.052	0.052	0.043	0.059	0.031	0.025	0.013	0.009	0.051
5%	0.015	0.011	0.027	0.010	0.101	0.212	0.029	0.275	0.184	0.039	0.028	0.106
10%	0.029	0.029	0.072	0.213	0.360	0.201	0.247	0.351	0.202	0.028	0.076	0.444
20%	0.107	0.066	0.069	0.280	0.459	0.366	0.312	0.399	0.833	0.079	0.082	0.502

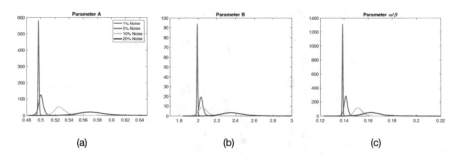

Fig. 8 Series of results showing how the posterior parameter distributions for the one-compartment model, with treatment and a low α/β ratio, depend on the level of noise in the data. As the noise level increases, the posteriors become less well-informed and tend to overestimate the parameter values compared to those used to generate the data

our one-compartment model requires only tumor volume data, whereas our two-compartment model also uses necrotic fraction data as an input, which is far more challenging and expensive to collect. Additionally, fitting the two-compartment model requires exploration of a five-dimensional parameter space, whereas fitting the one-compartment model requires investigation of a parameter space which is only three-dimensional. Therefore, we seek to understand when use of the one-compartment model is adequate, versus when the additional complexity of the two-compartment model (and cost for data collection) is necessary to accurately describe the tumor growth dynamics. In this section, we study the goodness of fit of the one-compartment model to synthetic data generated from the two-compartment model. In doing so, we address some of the questions raised here.

We generate synthetic data for total tumor volume and necrotic fraction from the two-compartment model, defined in Equations (4a) and (4b), by sweeping across the following region of five-dimensional parameter space: $\{(\lambda, K, \eta, \zeta, \beta) \in \mathbb{R}^5 \mid 0.2 \leq \lambda \leq 1, \ 0.1 \leq K \leq 1, \ 0 \leq \eta \leq \lambda, \ 0.5 \leq \zeta \leq 2, \ 0.014 \leq \beta \leq 0.14\}$. We generate a large number of samples ($O(10^3)$) using the Halton sequence, a quasi Monte Carlo method [13]. Then, we fit the synthetic tumor volume data using the one-compartment model defined in Equation (2).

We identify parameter regimes for which the one-compartment model provides a good approximation to the data from the two-compartment model and make note of other parameter regimes for which the one-compartment model does not approximate the two-compartment data well. The overall relation between the parameters are shown in Fig. 9; the parameter pairs (λ, A), (λ, B), are positively correlated since λ is the net tumor growth rate and, by definition, correlated to parameters representing the growth rate (A) and the inverse of the carrying capacity (B). Additionally, the values of β that represent radiosensitivity in the two models are positively correlated to each other, but not correlated to any other parameters, since the radiotherapy response is assumed to be independent of the tumor model parameters. We find negative correlation between the pair (η, A), since the natural death rate η is captured in the overall growth rate A, and between (K, B), also due

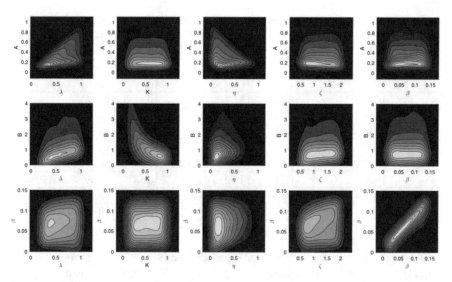

Fig. 9 Relationships between the two-compartment parameter samples $(\lambda, K, \eta, \zeta, \beta)$ and the fitted one-compartment parameters (A, B, β), depicted by contour plots showing the density of the parameter pairs. The parameter pairs (λ, A) and (λ, B) are positively correlated, while (η, A) and (K, B) are negatively correlated. The most apparent relation is between β from the two-compartment model and fitted β in the one-compartment model where the two values are linearly related

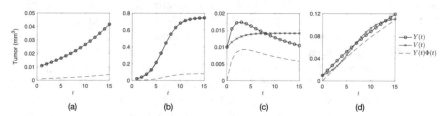

Fig. 10 Examples of tumor volume data $V(t)$ (○) and necrotic volume (- -) generated from the two-compartment model defined by Equations (4a) and (4b) and corresponding tumor volume values $Y(t)$ (×) obtained by fitting the one-compartment model defined by Equation (2) to the data. In subplots (**a**) and (**b**), the approximation is accurate; in subplots (**c**) and (**d**), the one-compartment model fails to accurately capture the behavior of the data. Although in (**d**), the overall trend is captured by the one-compartment model, the slope difference toward the end time point may lead to inaccurate predictions in future time points

to their inverse relation. The necrosis clearing rate ζ is not strongly correlated with any of the parameters from the one-compartment model.

Figure 10 displays simulation data generated from the two-compartment model in the absence of treatment. We plot the total tumor volume $(Y(0), Y(1), \cdots, Y(t_f))$ and the corresponding necrosis fraction $(\Phi(0), \Phi(1), \cdots, \Phi(t_f))$, where $t_f = 15$. Also shown are the corresponding fits $(V(0), V(1), \cdots, V(t_f))$ of the one-compartment model in Equation (2) to the synthetic data. We find that the

Table 4 Parameter values used to generate the plots in Figs. 10 and 12. They are the parameter values of the two-compartment model in Equations (4a) and (4b) and the fitted parameter values in the one-compartment model Equation (7)

	Two-comp. model					One-comp. model		
	λ	K	η	ζ	β	A	B	β
Fig. 10a	0.33	0.60	0.26	1.97		0.1159	2.4326	
Fig. 10b	0.89	0.90	0.23	1.90		0.6738	0.9061	
Fig. 10c	0.92	0.11	0.92	0.77		0.0834	4.9750	
Fig. 10d	0.954	0.29	0.92	0.03		0.3906	3.3525	
Fig. 12a	0.811	0.934	0.0278	1.58	0.0276	0.779	0.851	0.0123
Fig. 12b	0.508	0.705	0.115	1.50	0.102	0.382	0.59	0.0984
Fig. 12c	0.819	0.170	0.806	0.723	0.0234	0.113	5.00	0.0419
Fig. 12d	0.583	0.972	0.0205	0.531	0.127	0.499	0.468	0.1006

one-compartment model accurately fits synthetic data which is either monotonically increasing (Fig. 10a), or increasing and saturating (Fig. 10b). It is unable to accurately fit data generated from the two-compartment model for which the growth dynamics are either non-monotonic (Fig. 10c) or for which the necrotic region is large (Fig. 10d). We note that the poor fit of the one-compartment model to non-monotonic growth (Fig. 10c) is expected, since solutions to the one-compartment model are monotonic. The parameter values used to generate the synthetic data presented in Fig. 10 are included in Table 4, together with the parameter values obtained by fitting the one-compartment model to the synthetic data.

To quantify how well the one-compartment model fits data generated from the two-compartment model (i.e., the goodness of fit of the one-compartment model), we compute the relative error $e \doteq \|y - v\|_2 / \|y\|_2$ between the data $y = (Y(0), Y(1), \cdots, Y(t_f))$ and the fitted values $v = (V(0), V(1), \cdots, V(t_f))$ for each parameter sample. Each point plotted in Fig. 11 represents the relative error when the one-compartment model is fitted to synthetic data generated from the two-compartment model for a sample parameter set. Figure 11a shows how the goodness of fit of the one-compartment model decreases as the tumor size reduction indicator $(Y(t_f) / \max_t Y(t))$ decreases. Smaller values of this quantity correspond to tumors which have more pronounced, non-monotonic growth dynamics that cannot be modeled using the one-compartment model that only produces logistic curves. Figure 11b shows that the goodness of fit of the one-compartment model decreases as the necrotic volume of the synthetic data increases. This suggests that measuring the necrotic volume at a given time could be used to decide whether a one-compartment model can accurately fit the data or whether a two-compartment model is needed. More specifically, we observed an increased level of relative error e in some of the fits of the two-compartment data when the necrotic proportion $\Phi(t_f)$ at the final time point is large; the relative error can be large when the necrotic proportion $\Phi(t_f)$ is large.

We use a similar workflow to study the goodness of fit of the one-compartment model to the two-compartment model with radiotherapy. As in the case without

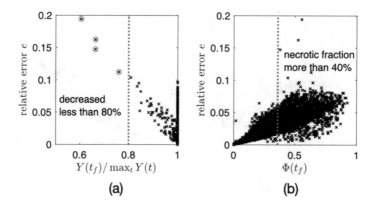

(a)   (b)

Fig. 11 Relative error e between the two-compartment tumor volume data generated from Equations (4a) and (4b) and the one-compartment fit using Equation (2) is plotted with respect to (**a**) tumor size reduction $Y(t_f)/\max_t Y(t)$ and (**b**) necrotic fraction $\Phi(t_f)$, where $t_f = 15$ is the final time of our simulation. Subplot (**a**) shows that data with non-monotonic tumor growth cannot be captured accurately by the one-compartment model. In particular, we highlight the data with $Y(t_f)/\max_t Y(t) < 0.8$ (o) that show large relative error values. Subplot (**b**) shows that the proportion of necrotic cells is also related to an increased error using the one-compartment model. We observe cases with increased error levels in data for which $\Phi(t_f) > 0.4$

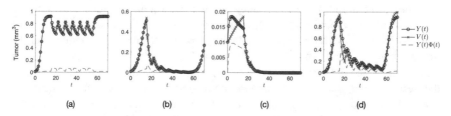

(a)   (b)   (c)   (d)

Fig. 12 Series of results showing the goodness of fit of the one-compartment model to data generated from the two-compartment model when treatment with radiotherapy is applied. We present synthetic data of tumor volume (o) and necrotic volume (- -) that were generated from the two-compartment model (see Equations (4a) and (4b)), together with tumor volume data (×) obtained by fitting to the one-compartment model (see Equation (7)). The fits are accurate in cases (**a**) and (**b**) but not in (**c**) or (**d**). In particular, the one-compartment model cannot reproduce the results of two-compartment model when the untreated growth dynamics are not well-fitted (**c**), and when the necrotic region is large (**d**)

treatment, Fig. 12 shows that there are situations for which the one-compartment model accurately captures the tumor dynamics (Fig. 12a, b) and others for which it does not (Fig. 12c, d). The two scenarios which typically yield inaccurate fits correspond to cases for which tumor growth before treatment can be not captured well (Fig. 12c) and/or the necrotic region is large (Fig. 12d). Although one can assume that the overall dynamics in Fig. 12d are captured reasonably well, the peak and trough during treatment could not be accurately fitted. This could potentially cause more inaccurate predictions when only a few, noisy data points are added. The parameter values used to generate the figure are included in Table 4.

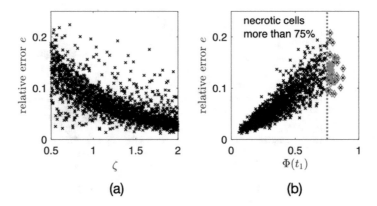

Fig. 13 Relative error e between the two-compartment data generated from Equations (4a) and (4b) with treatment and the one-compartment fit using Equation (7) with respect to (**a**) the parameter ζ and (**b**) necrotic fraction $\Phi(t_1)$, where $t_1 = 20$ is the time after the first week of treatment. The proportion of necrotic cells is positively correlated with the relative error. We highlight the data with $\Phi(t_f) > 0.75$ ($\Diamond$) that show large errors

As in the example without treatment, the relative error e is computed using the two-compartment tumor volume data y and the fitted tumor volume data from the one-compartment model v. In this case, the data is collected daily until the final time $t_f = 70$. In Fig. 13b, the error is plotted with respect to the necrotic fraction after the first week of treatment $\Phi(t_1)$. The results show that the relative error increases as the ratio of necrotic core increases. We observe a correlation of the relative error with the necrotic core, that is more apparent compared to the study without treatment (Fig. 11b). In addition, in Fig. 13a, we observe correlated patterns with the necrotic core decay rate ζ, as this is the parameter that determines the size of the necrotic core.

Next, we verify that the ratio of necrotic core $\Phi(t)$ is a robust indicator of goodness of fit of the one-compartment model with treatment fitted to synthetic data with noise. Figure 14 shows the relative error, e, against the necrotic fraction $\Phi(t)$, while increasing the noise level of the data up to 20%. A positively correlated relationship between the relative error of the one-compartment fit and the necrotic fraction is apparent once treatment is applied, while the noise in the data reduces the effectiveness of the indicator in cases without treatment. In general, larger noise in the data impacts the ability to accurately fit the one-compartment model to two-compartment data, as is shown in Fig. 15, where the fitted β in the one-compartment model becomes inaccurate as the noise increases. In conclusion, we determine that the necrotic fraction is a good indicator of the quality of fit of the one-compartment model to synthetic data generated from the two-compartment model.

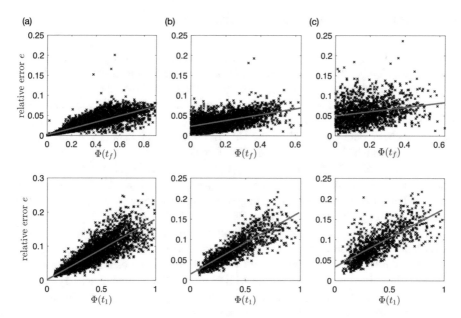

Fig. 14 The impact of adding noise to synthetic data from the two-compartment model on the ability to fit the one-compartment model without treatment (top) and with treatment (bottom). The relative error e is plotted with respect to the ratio of necrotic fraction $\Phi(t)$ while increasing noise in the data up to 20%. The least squares regression lines are shown. Despite the increased level of noise, the necrotic fraction remains a good indicator of the fitness of the one-compartment model when the treatment is given (bottom). (**a**) Noise 0%. (**b**) Noise 10%. (**c**) Noise 20%

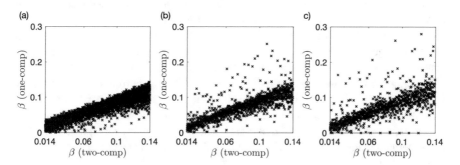

Fig. 15 The impact of adding noise to synthetic data from the two-compartment model reduces the ability to fit the one-compartment model. The parameter pairs of β that is used to generate the data from the two-compartment model and fitted using the one-compartment model show that larger noise in the data results in misfitted β values. (**a**) Noise 0%. (**b**) Noise 10%. (**c**) Noise 20%

5 Data Inclusion for Model Calibration

In an effort to predict how well simple mathematical models can be calibrated to clinical data of tumor volumes and to determine how much data is needed to accurately infer model parameters, we perform a sequential model calibration of: (i) the one-compartment model to data generated by the two-compartment model; (ii) the one-compartment model fit to CA data; and, (iii) the two-compartment model fit to synthetic data generated by the CA. In all cases, the data is generated from simulations in which tumors are treated with radiotherapy.

In each case, we fit the simpler model to a data set that includes tumor volumes, and necrotic fractions when relevant, measured once per week with 10% noise added. The calibration procedure is begun by fitting the lower-fidelity model to the first three data points (collected from the higher-fidelity model at days 8, 15, and 22); then the low-fidelity model is re-calibrated with the addition of each subsequent data point in an effort to determine a threshold at which we have "enough" data to accurately infer our model parameters with well-informed posterior distributions. When calibrating the one-compartment ODE model, the parameter set $[A, B, \beta]$ is estimated; for the two-compartment model, we estimate $[\lambda, K, \eta, \zeta, \beta]$. In all cases, α remains fixed at 0.14 to avoid identifiability issues.

At each calibration step, we calculate the relative error between the "fitted" and corresponding "true" parameter values. Since there is no explicit relationship between the parameters of the CA and those of the ODE models or mapping between parameters in the one- and two-compartment ODE models, we use a full set of in silico daily information to provide "true" parameter values. In each case, we initially fit the one- and two-compartment models to all 71 data points generated from the CA model and the one-compartment model to all 71 data points generated from the two-compartment model with no noise added and burn-in and subsequent MCMC chain lengths of 10,000. The parameter values generated from these fits are considered the "true" parameter values for each data set. We also assess the ability of the model to accurately predict the tumor regrowth (defined using the data point at day 70), by computing the absolute error in prediction when using the current calibrated model to predict forward in time. Figure 16 displays the calibration results when fitting to CA data in the "high" necrotic case (when the necrotic cells comprise 50% of total tumor volume in the absence of treatment). The final fits using all eight data points are generated using data with 10% noise added and burn-in and subsequent MCMC chain lengths of 10,000.

From Fig. 16, we see that when the necrotic fraction is high, the parameters converge well only when the two-compartment model is fit to both tumor volume and necrotic fraction data. In particular, the one-compartment model (Fig. 16a, b) does not accurately simulate a tumor with a high necrotic fraction; the fit is poor, and the estimated parameter values differ markedly from those used to generate the in silico data. Additionally, the entire fit changes when the final data point is added. This suggests that information about necrotic volume is needed to achieve a good fit to the data. When we fit the two-compartment model (Fig. 16c, d) to tumor

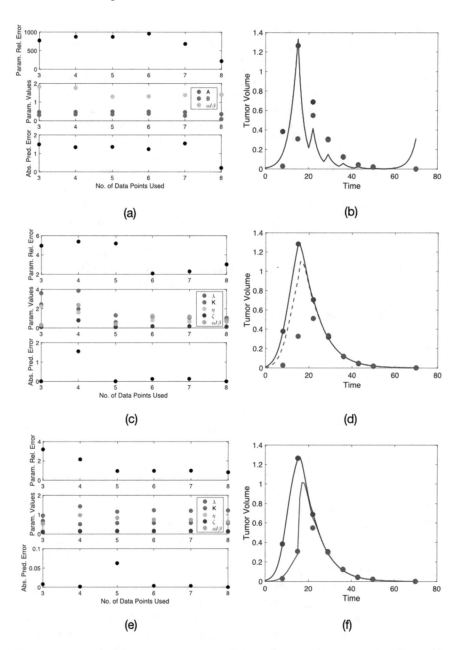

Fig. 16 Model fits with "high" necrotic fraction, "low" α/β ratio, and 10% noise. Blue represents tumor volume; red represents necrotic fraction. Solid curves are fit to the given data; dashed curves show necrotic fraction estimate from the two-compartment ODE model generated in absence of necrotic data. (**a**)–(**b**) One-compartment model fit, (**c**)–(**d**) Two-compartment model fit to tumor volume data only, (**e**)–(**f**) Two-compartment model fit to tumor volume and necrotic data generated from the CA

volume data only (no inclusion of necrotic information) generated from the CA, the overall fit is reasonable, but parameter values change markedly as the number of data points used for fitting varies. Additionally, the relative errors in the parameter values (as compared to those fit using the in silico data set) are high. Providing this supplemental information allows the slightly more complex two-compartment ODE model to accurately fit the high-necrotic tumor data; Fig. 16e, f shows a strong fit with greater consistency in the parameter estimates as data is added sequentially.

Analogous plots obtained by fitting CA data in the "low" necrotic case (when the necrotic cells comprise 20% of total tumor volume in the absence of treatment), are presented in Fig. 17. In this figure, all results relate to fits of the one-compartment model to the CA data as the tumor radiosensitivity varies. The radiosensitivity decreases from high ($\alpha/\beta = 1$) in (a)–(b), to medium ($\alpha/\beta = 3$) in (c)–(d), and low ($\alpha/\beta = 9$) in (e)–(f). Tumor regrowth data generated with medium and high α/β ratios is accurately predicted, even with only three data points. In the low α/β case, an additional data point is needed for accurate regrowth predictions. In all cases, the estimated parameter values converge rapidly to the true parameter values. We conclude that, despite its simplicity, for all three levels of radiosensitivity, the one-compartment model not only fits the data well, but can do so with just 3–4 few data points provided, for tumors with small necrotic regions.

6 Discussion

In this investigation, we have proposed a framework for choosing appropriate models, verifying the identifiability of their parameters, and calibrating those parameters with the available data. As a proof-of-concept, we investigated three distinct models of tumor growth: a one-compartment ODE model tracking tumor volume over time, a two-compartment model that includes an additional state variable representing the necrotic volume fraction, and a spatially explicit cellular automaton model that is more complex than the ODE models. We first showed that the one- and two-compartment model parameters are structurally identifiable without treatment, i.e. that one can uniquely recover all parameter values describing those models, given error-free model output data. However, we found that structural identifiability does not hold when treatment with radiotherapy is included in the models, since the radiosensitivity parameters are not uniquely identifiable. This led us to fix α and vary β only, when conducting model calibration.

Next we considered the practical identifiability of both ODE models. After adding varying levels of noise to synthetic tumor volume data, we found that the practical identifiability of parameter values becomes less well-informed for both the one-compartment and two-compartment models as the noise level increases. Further, the predicted parameter values tend to overestimate those values used to generate the data. By performing a sweep of the model parameters, we identified that the one-compartment model accurately fits synthetic data generated by the two-compartment model when the data is increasing monotonically or when the necrotic

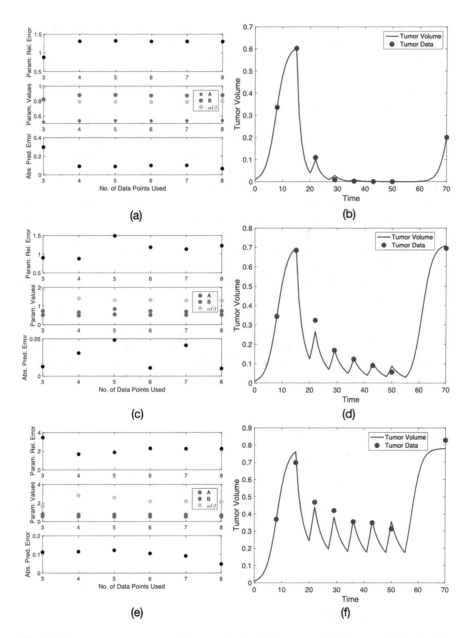

Fig. 17 The one-compartment model accurately describes tumors with a low necrotic fraction, for a range of α/β ratios. (**a**)–(**b**) $\alpha/\beta = 1$, (**c**)–(**d**) $\alpha/\beta = 3$, (**e**)–(**f**) $\alpha/\beta = 9$. In all three cases the relative errors in the estimated parameter values are small; the inclusion of the fourth data point results in absolute prediction error for tumor regrowth of less than 0.1 cm in all three cases

region is small. The error of the one-compartment fit to synthetic data increases as the size of the necrotic region increases, suggesting that in a clinical setting, the one-compartment ODE model becomes a less accurate predictor of tumor growth as the size of the necrotic region increases.

We also tested the ability of the simpler models to fit to data generated from the CA, using only a small number of data points. In agreement with the parameter sweep results, we found that for tumors with a large necrotic region, the estimated parameter values converge well only when fitting the two-compartment model to data on both total tumor and necrotic volumes. This implies that in cases with slow necrotic decay, information about tumor heterogeneity, rather than simply tumor volume, is necessary to fit these ODE models to the data. In such cases, i.e., when the necrotic region is large, the one-compartment ODE model will not accurately predict the response to treatment and tumor regrowth. However, when the necrotic fraction is small, it is possible to accurately identify parameters from the one-compartment model and to characterize tumor response to radiotherapy.

In the future, we aim to explore the sensitivity of the ODE models to intrinsic noise by adding noise to the parameter space before generating synthetic data and fitting the models to this data. We are also interested in incorporating multiple types of cells with different levels of radiosensitivity. We plan to examine how this additional complexity affects the overall predictive power of the ODE models and the amount of data needed to make accurate predictions. We also plan to conduct further experiments studying the impact of the quantity, type, and temporal location of available data on the ability to accurately calibrate various models.

Appendix: Structural Identifiability for the Two-Compartment Model

Below, we investigate the structural identifiability of the two-compartment model with radiotherapy, given in Equations (19) and (20), using the same techniques as presented in Sect. 3.1.

Case 1: No Radiation

In this case the model reads

$$
\frac{dY}{dt} = \lambda(1 - \Phi)Y(1 - (1 - \Phi)\frac{Y}{K}) - \xi\Phi Y
$$

$$
\frac{d\Phi}{dt} = (1 - \Phi)\left[\eta - \lambda\Phi(1 - (1 - \Phi)\frac{Y}{K} - \xi\Phi)\right]
$$

with unknown parameters $p = \{\lambda, K, \xi, \eta\}$, observable quantities $y(p; t) = \{Y, \Phi\}$, and known initial conditions. We repeat the analysis as before, using the Taylor coefficients. We define the following known quantities:

$$a_0 = Y(0^+) \quad b_0 = \phi(0^+) \quad a_1 = Y'(0^+) \quad b_1 = \phi'(0^+) \quad a_2 = Y''(0^+) \quad b_2 = \phi''(0^+).$$

We substitute these quantities into the model system to obtain:

$$a_1 = \lambda(1 - b_0)a_0[1 - (1 - b_0)\frac{a_0}{K}] - \xi b_0 a_0$$

$$= [-(1 - b_0)^2 a_0^2]\frac{\lambda}{K} + [(1 - b_0)a_0]\lambda - [b_0 a_0]\xi$$

$$b_1 = (1 - b_0)\left[\eta - \lambda b_0(1 - (1 - b_0)\frac{a_0}{K} - \xi b_0)\right]$$

$$+ [-(1 - b_0)b_0]\lambda + [(1 - b_0)^2 b_0 a_0]\frac{\lambda}{K} - [(1 - b_0)b_0]\xi + [(1 - b_0)]\eta.$$

We differentiate the model equations once more to obtain:

$$a_2 = \frac{(-2\lambda(-1 + b_0)^2 a_0 - ((\xi + \lambda)b_0 - \lambda)K)a_1 - a_0 b_1(2(-1 + b_0)\lambda a_0 + (\xi + \lambda)K)}{K}$$

$$b_2 = \frac{(3a_0 b_0^2 \lambda + (-4\lambda a_0 + 2(\xi + \lambda)K)b_0 + \lambda a_0 - K(\xi + \eta + \lambda))b_1 + b_0 a_1 \lambda(-1 + b0)^2)}{K}.$$

The above four equations can now be used to solve for each parameter as follows:

$$K = \frac{1}{[((b_0 - 1)b_2 - b_1^2)a_0^2 + a_2(b_0 - 1)^2 a_0 - a_1^2(b_0 - 1)^2]b_1}$$

$$\times [(b_1(b_0 + 1)(b_0 b_2 - b_1^2 - b_2)a_0^3 + ((a_1 b_2 + a_2 b_1)b_0^2 + ((-b_1^2 - b_2)a_1 + a_2 b_1)b_0$$

$$- a_1 b_1^2 - a_2 b_1)(-1 + b0)a_0^2 - (-a_2 b_0^3 + (a_1 b_1 + a_2)b_0^2 + 2a_1 b_1 b_0 - 2a_1 b_1)$$

$$\times a_1(-1 + b_0)a_0 - a_1^3 b_0^2(-1 + b_0)^2)(-1 + b_0)],$$

$$\lambda = \frac{-1}{(a_0 b_0 b_1 + a_1 b_0^2 - a_0 b_1 - 2a_1 b_0 + a_1)a_0^2 b_1}$$

$$\times [a_0^3 b_0^3 b_1 b_2 - a_0^3 b_0 b_1^3 + a_0^2 a_1 b_0^3 b_2 - a_0^2 a_1 b_0^2 b_1^2$$

$$+ a_0^2 a_2 b_0^3 b_1 - a_0 a_1^2 b_0^3 b_1 + a_0 a_1 a_2 b_0^4 - a_1^3 b_0^4 - a_0^3 b_1^3$$

$$- 2a_0^2 a_1 b_0^2 b_2 - a_0 a_1^2 b_0^2 b_1 - 2a_0 a_1 a_2 b_0^3$$

$$+ 2a_1^3 b_0^3 - a_0^3 b_1 b_2 + a_0^2 a_1 b_0 b_2 + a_0^2 a_1 b_1^2 - 2a_0^2 a_2 b_0 b_1 + 4a_0 a_1^2 b_0 b_1$$

$$+ a_0 a_1 a_2 b_0^2 - a_1^3 b_0^2 + a_0^2 a_2 b_1 - 2a_0 a_1^2 b_1]$$

$$\eta = -\frac{a_0^2 b_0 b_2 - a_0^2 b_1^2 + a_0 a_2 b_0^2 - a_1^2 b_0^2}{b_1 a_0^2}$$

$$\xi = \frac{a_0^2 b_0 b_2 - a_0^2 b_1^2 + a_0 a_2 b_0^2 - a_1^2 b_0^2 - a_0^2 b_2 - a_0 a_1 b_1 - a_0 a_2 b_0 + a_1^2 b_0}{b_1 a_0^2}$$

Since we are able to obtain unique solutions for each of the four parameters, we declare them to be structurally identifiable. For GenSSI, only two Lie derivatives are needed which yield rank 4, and thus results show all four parameters are structurally identifiable, in agreement with our calculations above.

In addition to the above analysis, we also repeated the analysis in the case in which only the tumor volume could be observed, (i.e., $y(t; p) = Y(t)$), but with known initial conditions in tumor volume and necrotic fraction. In this case, we took higher order Taylor series coefficients (up to order 4) and obtained that $p = \{\lambda, K, \xi, \eta\}$ were structurally identifiable. Similarly, GenSSI took Lie derivatives up to order 4 and confirmed that all parameters were structurally identifiable.

Case 2: With Radiation Treatment

Similar to the single compartment model, here we examine the effect of a point treatment. The model equations read:

$$\frac{dY}{dt} = \lambda(1 - \Phi)Y(1 - (1 - \Phi)\frac{Y}{K}) - \xi \Phi Y$$

$$\frac{d\Phi}{dt} = (1 - \Phi)\left[\eta - \lambda\Phi(1 - (1 - \Phi)\frac{Y}{K}) - \xi\Phi\right], \quad \text{for} \quad t_i^+ < t < t_{i+1}^-$$

$$\Phi(t_i^+) = \Phi(t_i^-) + (1 - \Phi(t_i^-))(1 - \Gamma),$$

where $\Gamma = \exp(-\alpha d - \beta d^2)$. Since the other parameters are known and measured prior to treatment, as in the previous section, we want to solve for $p = \{\alpha, \beta\}$ assuming $y(p; t) = \{Y, \Phi\}$ as observable quantities. We let

$$A_0 = Y(t_i^+) \quad B_0 = \Phi(t_i^-) \quad A_1 = Y'(t_i^+) \quad B_1 = \Phi'(t_i^+),$$

and substitute these quantities into the model equations:

$$A_1 = \frac{-A_0^2 \lambda (B_0 - 1)^2}{K} \Gamma^2 - A_0(\xi + \lambda)(B_0 - 1)\Gamma - A_0 \xi$$

$$B_1 = \frac{[A_0 \lambda (B_0 - 1)^2 \Gamma^2 + (B_0 - 1)((K + A_0)\lambda + \xi K)\Gamma + K(\xi - \eta + \lambda)]\Gamma(B_0 - 1)}{K}$$

As with the one-compartment model, we find that the equations are not informative for α and β simultaneously, thus, we again declare the pair (α, β) to be non-identifiable in this setting. As before, we choose to fix α for all subsequent model calibrations and measure the ratio α/β to use as a measure of radiosensitivity.

References

1. E. Balsa-Canto, A. A. Alonso, J. R. Banga. An iterative identification procedure for dynamic modeling of biochemical networks. BMS Systems Biology. **4(11)**, (2010). https://doi.org/10.1186/1752-0509-4-11.
2. R. Bellman, K. J. Astrom. On structural Identifiability. Mathematical Biosciences, **7**, 329–339 (1970).
3. M.A. Boemo, H.M. Byrne. Mathematical modelling of a hypoxia-regulated oncolytic virus delivered by tumour-associated macrophages. J Theor Biol **461**, 102–116 (2019).
4. H. Byrne, L. Preziosi (2003). Modelling solid tumour growth using the theory of mixtures, Math Med Biol **20**, 341–366.
5. M. J. Chappell, K. R. Godfrey, S. Vajda. Global identifiability of the parameters of nonlinear systems with specified inputs: a comparison of methods. Mathematical Biosciences, **102**, 41–73 (1990).
6. O.T. Chis, J.R. Banga, E. Balsa-Canto. Structural identifiability of systems biology models: a critical comparison of methods. PLOS One. **6(11)**, 1–16 (2011). https://doi.org/10.1371/journal.pone.0027755.
7. O.T. Chis, J.R. Banga, E. Balsa-Canto. GenSSI: a software toolbox for structural identifiability analysis of biological models. Bioinformatics. **27(18)**, 2610–2611 (2011). https://doi.org/10.1093/bioinformatics/btr431.
8. J. Collis, A.J. Connor, M. Paczkowski, P. Kannan, J. Pitt-Francis, H.M. Byrne, M.E. Hubbard. Bayesian calibration, validation and uncertainty quantification for predictive modeling of tumor growth: a tutorial. Bull. Math. Biol. **79**(4), 939–974. (2017).
9. J.M.J. da Costa, H.R.B. Orlande, W.B. da Silaa. Model selection and parameter estimation in tumor growth models using approximate Bayesian computation – ABC. Comp. Appl. Math. **37**(3), 2795–2815. (2018).
10. H. Enderling, M.A.J. Chaplain, P. Hahnfeldt. Quantitative modeling of tumor dynamics and radiotherapy. Acta Biotheoretica. **58**(4), 341–353. (2010).
11. H. Haario, M. Laine, A. Mira, et al.: Efficient adaptive MCMC. Stat. Comput. **26**, 339–354 (2006).
12. E.J. Hall. Radiobiology for the radiologist. J.B. Lippincott, Philadelphia, 478–480 (1994).
13. N. Harald. Random number generation and quasi-Monte Carlo method. SIAM (1992).
14. P. Kannan, M. Paczkowski, A. Miar, et al.: Radiation resistant cancer cells enhance the survival and resistance of sensitive cells in prostate spheroids. bioRxiv (2019). https://doi.org/10.1101/564724.
15. J. Kursawe, R.E. Baker, A.G. Fletcher. Approximate Bayesian computation reveals the importance of repeated measurements for parameterising cell-based models of growing tissues. J. Theor. Biol. **443**, 66–81. (2018)
16. B. Lambert, A.L. MacLean, A.G. Fletcher, A.N. Combes, M.H. Little, H.M. Byrne. Bayesian inference of agent-based models: a tool for studying kidney branching morphogenesis. J. Math. Biol. **76**(7), 1673–1697. (2018).
17. D.E. Lea, D.G. Catcheside. The mechanism of the induction by radiation of chromosome aberrations in tradescantia. Journal of Genetics. **44**, 216–245 (1942).
18. T.D. Lewin. Modelling the impact of heterogeneity in tumor composition on the response to fractionated radiotherapy. D. Phil. Thesis, University of Oxford, 2018.

19. T.D. Lewin, H.M. Byrne, P.K. Maini, J.J. Caudell, E.G. Moros, H. Enderling. The importance of dead material within a tumour on the dynamics in response to radiotherapy. Physics in Medicine and Biology. https://doi.org/10.1088/1361-6560/ab4c27 (2019).

20. T.D. Lewin, P.K. Maini, E.G. Moros, H. Enderling, H.M. Byrne. A three-phase model to investigate the effects of dead material on the growth of avascular tumours. Mathematical Modelling of Natural Phenomena (in press) (2019).

21. E. Lima, J.T. Oden, D.A. Hormuth 2nd, T.E. Yankeelov, R.C. Almeida. Selection, calibration, and validation of models of tumor growth. Mathematical Models and Methods in Applied Sciences 26(12), 2341–2368. (2016).

22. J.T. Oden, A. Hawkins, S. Prudhomme. General diffuse-interface theories and an approach to predictive tumour growth modelling. Math. Models Meth. Appl. Sci. 20 (3), 477–517 (2010).

23. H. Pohjanpalo. System identifiability based on the power series expansion of the solution. Mathematical Biosciences. 41, 21–33 (1978).

24. B. Ribba, N.H. Holford, P. Magni, I Troconiz, I Gueorguieva, P. Girard, C. Sarr, M. Elishmereni, C. Kloft, L.E. Friberg. A review of mixed-effects models of tumor growth and effects of anti-cancer treatment used in population analysis. CPT Pharmacometrics Syst. Pharmacol. 3, e113. (2014).

25. R.C. Smith. Uncertainty Quantification: Theory, Implementation, and Applications. SIAM Computational Science and Engineering Series (CS12). (2014).

Correction to: Investigating the Impact of Combination Phage and Antibiotic Therapy: A Modeling Study

Selenne Banuelos, Hayriye Gulbudak, Mary Ann Horn, Qimin Huang, Aadrita Nandi, Hwayeon Ryu, and Rebecca Segal

Correction to:
Chapter 6 in: R. Segal et al. (eds.),
***Using Mathematics to Understand Biological Complexity*,**
Association for Women in Mathematics Series 22,
https://doi.org/10.1007/978-3-030-57129-0_6

The original version of this chapter was inadvertently published without updating additional corrections from the author. Now, the corrections have been incorporated in chapter proof and front matter.

The updated online version of this chapter can be found at
https://doi.org/10.1007/978-3-030-57129-0_6

Printed in the United States
by Baker & Taylor Publisher Services